World of Computing

Gerard O'Regan

World of Computing

A Primer Companion for the Digital Age

 Springer

Gerard O'Regan
SQC Consulting
Mallow, Cork
Ireland

ISBN 978-3-319-75843-5   ISBN 978-3-319-75844-2   (eBook)
https://doi.org/10.1007/978-3-319-75844-2

Library of Congress Control Number: 2018932550

Printed on acid-free paper

This Springer imprint is published by the registered company Springer International Publishing AG
part of Springer Nature
The registered company address is: Gewerbestrasse 11, 6330 Cham, Switzerland

To
Kevin Crowley

For lifelong friendship

Preface

Overview

The objective of this book is to provide a concise introduction to the world of computing to students and general readers. The computing field is a vast area and so it is not possible to cover every area of computing, or to cover every topic in detail. Therefore, the goal is to give the reader a flavour of some of the important areas of computing, and to stimulate the reader to study the more advanced articles and books that are available.

Organisation and Features

Chapter 1 introduces analog and digital computers, and the von Neumann architecture which is the fundamental architecture underlying a digital computer. Chapter 2 discusses the foundations of computing, and we describe the binary number system and the Step Reckoner calculating machine that were invented by Leibniz. Babbage designed the Difference Engine as a machine to evaluate polynomials, and his Analytic Engine provided the vision of a modern computer. Boole was an English mathematician who made important contributions to mathematics and logic, and his symbolic logic is the foundation for digital computing.

Chapter 3 presents a concise history of computing including a discussion of the first digital computers, the first commercial computers, the SAGE air defence system, the invention of the transistor at Bell Labs, and early transistor computers, the invention of the integrated circuit at Texas Instruments, the development of the IBM System/360, and its influence on later computer development, later mainframes and minicomputers, including DEC's minicomputers. We discuss the revolutionary invention of the microprocessor, and how it led to the development of home computers such as the Apple I and II, Commodore computers, Atari computers, the Sinclair ZX 81 and ZX spectrum computers and the Apple Macintosh.

We discuss the introduction of the IBM personal computer, which was a major milestone in the computing field.

Chapter 4 introduces the essential mathematics for computing including sets, relations and functions. Sets are collections of well-defined objects; relations indicate relationships between members of two sets A and B; and functions are a special type of relation where there is exactly (or at most) one relationship for each element $a \in A$ with an element in B.

Chapter 5 presents a short introduction to algorithms, where an algorithm is a well-defined procedure for solving a problem. It consists of a sequence of steps that takes a set of values as input, and produces a set of values as output. An algorithm is an exact specification of how to solve the problem, and it explicitly defines the procedure so that a computer program may implement the solution in some programming language.

Chapter 6 presents an introduction to logic for computing, and includes a short history of logic, and an introduction to propositional and predicate logic. Propositional logic is the study of propositions, where a proposition is a statement that is either true or false. It may be used to encode simple arguments that are expressed in natural language, and to determine their validity. Predicate logic allows complex facts about the world to be represented, and new facts may be determined via deductive reasoning.

Chapter 7 discusses human–computer interaction (HCI), which is a branch of computer science that is concerned with the design, evaluation and implementation of interactive computing systems for human use. It is focused on the interfaces between people and computers, and involves several fields including computer science, cognitive psychology, design and communication.

Chapter 8 presents a short introduction to programming languages starting with machine languages; to assembly languages; to early high-level procedural languages such as Fortran and COBOL; to later high-level procedural languages such as Pascal and C; and to object-oriented languages such as C++ and Java. Functional programming languages and logic programming languages are discussed, and there is a short discussion on the important area of syntax and semantics.

Chapter 9 presents a short introduction of the software engineering field. We discuss its key challenges and several high-profile software failures. The waterfall and spiral lifecycles are discussed, as well as a brief discussion on the rational unified process and the popular Agile methodology. We discuss the key activities in the waterfall model such as requirements, design, implementation, unit, system and acceptance testing.

Chapter 10 presents a short introduction to operating systems including the IBM OS/360, which was the operating system for the IBM System/360 family of computers. We discuss the MVS and VM operating systems, which were used on the IBM System/370 mainframe computer. We discuss the UNIX operating system, which is a multi-user, and multi-tasking operating system. It was written almost entirely in C. DEC developed the VAX/VMS operating system in the late 1970s for its VAX family of minicomputers. Microsoft developed MS/DOS for the IBM personal computer in 1981, and it introduced Windows as a response to the

GUI-driven operating system of the Apple Macintosh. There is a short discussion on Android and iOS, which are popular operating systems for mobile devices.

Chapter 11 presents a short introduction to databases including a discussion of the hierarchical and network models. We discuss the relational database model which was developed by Codd at IBM in more detail, as most databases used today are relational. There is a short discussion on the SQL query language and on the Oracle database.

Chapter 12 presents a short introduction to telecommunications, and it focuses on the development of mobile phone technology. The development of the AXE system by Ericsson is discussed, as this was the first fully automated digital switching system. We discuss the concept of a cellular system, which was introduced by Bell Labs, as well as the introduction of the first mobile phone, the DynaTAC, by Motorola.

Chapter 13 describes the Internet revolution starting from ARPANET, which was a packet switched network and TCP/IP, which is a set of network standards for interconnecting networks and computers. These developments led to the birth of the Internet, and Tim Berners-Lee's work at CERN led to the birth of the World Wide Web. We discuss various applications of the World Wide Web, as well as the dot-com bubble and burst of the late 1990s/early 2000. We conclude with a brief discussion on the Internet of Things and the Internet of Money (including bitcoin).

Chapter 14 discusses the invention of the smartphone and the rise of social media. It describes the evolution of the smartphone from PDAs and mobile phone technology, and the impact of Facebook and Twitter in social networking is discussed. Facebook has become a way for young people to discuss their hopes and aspirations as well as a tool for social protest and revolution. Twitter has become a popular tool in political communication, and it is also an effective way for businesses to advertise its brand to its target audience. We briefly discuss how social media has been abused by individuals and states to spread fake news, and the challenges this poses to western democracies.

Chapter 15 discusses legal aspects of computing and is concerned with the overlap of the law and the computing field. We discuss intellectual property such as patents, copyright and trademarks and the licensing of software. We examine the area of hacking and computer crime, and explore the nature of privacy, free speech and censorship. We consider the legal issues of bespoke software development, and the legal aspects of the Internet.

Chapter 16 discusses ethics and professional responsibility in computing. Ethics is a branch of philosophy that deals with moral questions such as what is right or wrong, and computer ethics are a set of principles that guide the behaviour of individuals when using computer resources. Professional ethics are a code of conduct that governs how members of a profession deal with each other and with third parties.

Chapter 17 discusses innovation in the computing field including distributed systems, service-oriented architecture (SOA), Software as a Service (SaaS), cloud computing and embedded systems. Chapter 18 is the concluding chapter in which we summarise the journey that we have travelled in the book.

Audience

The main audience of this book are computer science students and the general reader who are interested in learning about the fascinating world of computing.

Acknowledgements

I am indebted to friends and family who supported my efforts in this endeavour. This book is dedicated to Kevin Crowley (a lifelong friend) who is currently fighting a difficult illness with all his strength and determination. I wish him victory in this major challenge, and my thoughts are with him and his family at this difficult time. I would like to thank the team at Springer for their constant professional work and support. I would like to thank all copyright owners for the permission to use their images. I believe that all of the required permissions have been obtained, but in the unlikely event that an image has been used without the appropriate authorisation, please contact me so that the required permission can be obtained.

Cork, Ireland   Gerard O'Regan

Contents

List of Figures

List of Tables

Chapter 1
What Is a Computer?

1.1 Introduction

It is difficult to think of western society today without modern technology. We have witnessed in recent decades a proliferation of high-tech computers, mobile phones, text messaging, the Internet, the World Wide Web and social media. Software is pervasive and it is an integral part of automobiles, airplanes, televisions and mobile communication. The pace of change is relentless, and communication today is instantaneous with technologies such as Skype, Twitter and WhatsApp.

Today, people may book flights over the World Wide Web as well as keeping in contact with friends and family members around the world. In previous generations, communication involved writing letters that often took months to reach the recipient. However, today's technology has transformed the modern world into a global village, and the modern citizen may make video calls over the Internet or post pictures and videos on social media sites such as Facebook and Twitter. The World Wide Web allows business to compete in a global market.

A computer is a programmable electronic device that can process, store and retrieve data. It processes data according to a set of instructions or program. All computers consist of two basic parts namely *hardware* and *software*. The hardware is the physical part of the machine, and the components of a digital computer include memory for short-term storage of data or instructions; an arithmetic/logic

© Springer International Publishing AG, part of Springer Nature 2018

G. O'Regan, *World of Computing*,

https://doi.org/10.1007/978-3-319-75844-2_1

unit for carrying out arithmetic and logical operations; a control unit responsible for the execution of computer instructions in memory; and peripherals that handle the input and output operations. Software is a set of instructions that tells the computer what to do.

The original meaning of the word '*computer*' referred to someone who carried out calculations rather than an actual machine. The early digital computers built in the 1940s and 1950s were enormous machines consisting of thousands of vacuum tubes. They typically filled a large room but their computational power was a fraction of the personal computers and mobile devices used today.

There are two distinct families of computing devices namely *digital computers* and the historical *analog computer*. The earliest computers were analog not digital, and these two types of computer operate on quite different principles.

The computation in a digital computer is based on binary digits: i.e. '0' and '1'. Electronic circuits are used to represent binary numbers, with the state of an electrical switch (i.e. 'on' or 'off') representing a binary digit internally within a computer.

A digital computer is a sequential device that generally operates on data one step at a time, and the earliest digital computers were developed in the 1940s. The data are represented in binary format, and a single transistor (initially bulky vacuum tubes) is used to represent a binary digit. Several transistors are required to store larger numbers.

An *analog computer* operates in a completely different way to a digital computer. The representation of data in an analog computer reflects the properties of the data that are being modelled. For example, data and numbers may be represented by physical quantities such as electric voltage, whereas a stream of binary digits is used to represent them in a digital computer.

1.2 Analog Computers

James Thompson (who was the brother of the physicist Lord Kelvin) did early foundational work on analog computation in the nineteenth century. He invented a wheel-and-disk integrator, which was used in mechanical analog devices, and he worked with Kelvin to construct a device to perform the integration of a product of two functions. Kelvin later described a general-purpose analog machine (he did not build it) for integrating linear differential equations. He built a tide predicting analog computer that remained in use at the Port of Liverpool up to the 1960s.

The operations in an analog computer are performed in parallel, and they are useful in simulating dynamic systems. They have been applied to flight simulation, nuclear power plants and industrial chemical processes.

Vannevar Bush at the Massachusetts Institute of Technology developed the first large-scale general-purpose mechanical analog computer. Bush's Differential Analyser (Fig. 1.1) was a mechanical analog computer designed to solve sixth-order differential equations by integration, using wheel-and-disk mechanisms

Fig. 1.1 Vannevar Bush with the Differential Analyser

to perform the integration. The mechanisation allowed integration and differential equations problems to be solved more rapidly. The machine took up the space of a large table in a room and weighed about 100 tonnes.

It contained wheels, disks, shafts and gears to perform the calculations. It required a considerable set-up time by technicians to solve an equation. It contained 150 motors and miles of wires connecting relays and vacuum tubes.

Data representation in an analog computer is compact, but it may be subject to corruption with noise. A single capacitor can represent one continuous variable in an analog computer. Analog computers were replaced by digital computers shortly after the Second World War.

1.3 Digital Computers

Early digital computers used vacuum tubes to store binary information, and a vacuum tube may represent the binary value '0' or '1'. These tubes were large and bulky and generated a significant amount of heat. Air conditioning was required to cool the machine, and there were problems with the reliability of the tubes.

Shockley and others invented the transistor in the late 1940s, and it replaced vacuum tubes from the late 1950s onwards. Transistors are small and consume very little power, and the resulting machines were smaller, faster and more reliable.

Integrated circuits were introduced in the early 1960s, and a massive amount of computational power could now be placed on a very small chip. Integrated circuits are small and consume very little power, and may be mass-produced to a very high-quality standard. However, integrated circuits are difficult to modify or repair, and are nearly always replaced on failure.

The fundamental architecture of a computer has remained basically the same since von Neumann and others proposed it in the mid-1940s. It includes a central processing unit which includes the control unit and the arithmetic unit, an input and output unit and memory.

1.3.1 Vacuum Tubes

A vacuum tube is a device that relies on the flow of an electric current through a vacuum. Vacuum tubes (*thermionic valves*) were widely used in electronic devices such as televisions, radios and computers until the invention of the transistor.

The basic idea of a vacuum tube is that the current passes through the filament, which then heats it up so that it gives off electrons. The electrons are negatively charged and are attracted to the small positive plate (or anode) within the tube. A unidirectional flow is thus established between the filament and the plate. Thomas Edison had observed this while investigating the reason for breakage of lamp filaments. He noted an uneven blackening (darkest near one terminal of the filament) of the bulbs in his incandescent lamps, and noted that current flows from the lamp's filament and a plate within the vacuum.

The first generation of computers used several thousand bulky vacuum tubes, with several racks of vacuum tubes taking up the space of a large room. The vacuum tube used in the early computers was a three-terminal device, and it consisted of a cathode, a grid and a plate. It was used to represent one of two binary states: i.e. the binary value '0' or '1'.

The filament of a vacuum tube becomes unstable over time. In addition, if air leaks into the tube then oxygen will react with the hot filament and damage it. The size and unreliability of vacuum tubes motivated research into more compact and reliable technologies. This led to the invention of the transistor in the late 1940s.

The first generation of digital computers all used vacuum tubes: e.g. the Atanasoff–Berry Computer (ABC) developed at the University of Iowa in 1942; Colossus developed at Bletchley Park in 1944; and ENIAC developed in the United States in the mid-1940s.

1.3.2 Transistors

The transistor is a fundamental building block in modern electronic systems, and its invention revolutionised the field of electronics. It was smaller, cheaper and more reliable than the existing vacuum tubes.

The transistor is a three-terminal, solid-state electronic device. It can control electric current or voltage between two of the terminals by applying an electric current or voltage to the third terminal. The three-terminal transistor enables an electric switch to be made which can be controlled by another electrical switch.

Complicated logic circuits may be built up by cascading these switches (switches that control switches that control switches, and so on).

These logic circuits may be built very compactly on a silicon chip with a density of over a million transistors per square centimetre. The switches may be turned on and off very rapidly (e.g. every 0.000000001 s). These electronic chips are at the heart of modern electronic devices.

The transistor (Fig. 1.2) was developed at Bell Labs after the Second World War. The goal of the research was to find a solid-state alternative to vacuum tubes, as this technology was too bulky and unreliable. Three Bell Labs inventors (Shockley, Bardeen and Brattain) were awarded the Nobel Prize in physics in 1956 in recognition of their invention of the transistor.

William Shockley (Fig. 3.14) was involved in radar research and anti-submarine operations research during the Second World War, and after the war he led the Bell Labs research group (that included Bardeen and Brattain) that aimed to find a solid-state alternative to the glass-based vacuum tubes.

Bardeen and Brattain succeeded in creating a point contact transistor in 1947 independently of Shockley who was working on a junction-based transistor. Shockley believed that the points contact transistor would not be commercially viable, and his junction point transistor was announced in 1951.

Shockley formed Shockley Semiconductor Inc. (part of Beckman Instruments) in 1955. The second generation of computers used transistors instead of vacuum tubes. The University of Manchester's experimental Transistor Computer was one of the earliest transistor computers. The prototype machine appeared in 1953 and the full-size version was commissioned in 1955. The invention of the transistor is discussed in more detail in Chap. 3.

Fig. 1.2 Replica of transistor. Courtesy of Lucent Bell Labs

1.3.3   Integrated Circuits

Jack Kilby (Fig. 3.16) of Texas Instruments invented the integrated circuit in 1958. His invention used a wafer of germanium, and Robert Noyce of Fairchild Semiconductors did subsequent work on silicon-based integrated circuits. The integrated circuit was an effective solution to the problem of building a circuit with many components, and the Nobel Prize in Physics was awarded to Kirby in 2000 for his contributions to its invention.

An integrated circuit consists of a set of electronic circuits on a small chip of semiconductor material, and it is much smaller than a circuit made from independent components. Integrated circuits today are extremely compact, and may contain billions of transistors and other electronic components in a tiny area. The width of each conducting line has got smaller and smaller over the years due to advances in technology, and it is now measured is in tens of nanometres.

The number of transistors per unit area has been doubling (roughly) every 1–2 years over the last 30 years. This amazing progress in circuit fabrication is known as Moore's law after Gordon Moore (one of the founders of Intel) who formulated the law in the mid-1960s (O'Regan 2013).

Kilby was designing micro-modules for the military and this involved connecting many germanium[1] wafers of discrete components together by stacking each wafer on top of one another. The connections were made by running wires up the sides of the wafers.

Kilby saw this process as unnecessarily complicated and realised that if a piece of germanium was engineered properly, it could act as many components simultaneously. That is, instead of making transistors one-by-one, several transistors could be made at the same time on the same piece of semiconductor. In other words, transistors and other electric components such as resistors, capacitors and diodes can be made by the same process with the same materials.

This idea led to the birth of the first integrated circuit and its development involved miniaturising transistors and placing them on silicon chips called semiconductors. The use of semiconductors led to third generation computers, with a major increase in speed and efficiency.

Users interacted with third generation computers through keyboards and monitors and interfaced with an operating system, which allowed the device to run several applications at a time with a central program that monitored the memory. Computers became accessible to a wider audience, as they were smaller and cheaper than their predecessors. The invention of the integrated circuit is discussed in more detail in Chap. 3.

---

[1]Germanium is an important semiconductor material used in transistors and other electronic devices.

1.3.4 Microprocessor S

The Intel P4004 microprocessor (Fig. 3.20) was the world's first microprocessor, and it was released in 1971. It was the first semiconductor device that provided, at the chip level, the functions of a computer.

The invention of the microprocessor happened by accident rather than design. Busicom, a Japanese company, requested Intel to design a set of integrated circuits for its new family of high-performance programmable calculators. Ted Hoff, an Intel engineer, studied Busicom's design and rejected it as unwieldy. He proposed a more elegant solution requiring just four integrated circuits (Busicom's required 12 integrated circuits), and his design included a chip that was a general-purpose logic device that derived its application instructions from the semiconductor memory. This was the Intel 4004 microprocessor.

It provided the basic building blocks that are used in today's microcomputers, including the arithmetic and logic unit and the control unit. The 4-bit Intel 4004 ran at a clock speed of 108 kHz and contained 2300 transistors. It processed data in four bits, but its instructions were eight bits long. It could address up to 1 Kb of program memory and up to 4 Kb of data memory.

Gary Kildall of Digital Research was one of the early people to recognise the potential of a microprocessor as a computer. He worked as a consultant with Intel, and he began writing experimental programs for the Intel 4004 microprocessor. He later developed the CP/M operating system for the Intel 8080 chip, and he set up Digital Research to commercialise the operating system.

The development of the microprocessor led to the fourth generation of computers with thousands of integrated circuits placed onto a single silicon chip. A single chip could now contain all the components of a computer from the CPU and memory to input and output controls. It could fit in the palm of the hand whereas first generation of computers filled an entire room. The invention of the microprocessor is discussed in more detail in Chap. 3.

1.4 Von Neumann Architecture

The earliest computers were fixed program machines that were designed to do a specific task. This proved to be a major limitation as it meant that a complex manual rewiring process was required to enable the machine to solve a different problem.

The computers used today are general-purpose machines designed to allow a variety of programs to be run on the machine. Von Neumann and others (von Neumann 1945) described the fundamental architecture underlying the computers used today in the late 1940s. It is known as von Neumann architecture (Fig. 1.3).

The von Neumann architecture arose on work done by von Neumann, Eckert, Mauchly and others on the design of the EDVAC computer (which was the

Fig. 1.3 Von Neumann
architecture

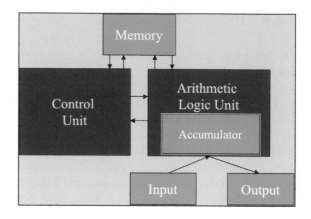

Fig. 1.4 Fetch/Execute cycle

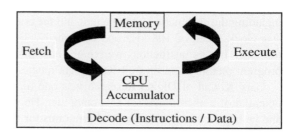

successor to ENIAC computer). Von Neumann's draft report on EDVAC (von Neumann 1945) described the new architecture (Table 1.1).

The architecture led to the birth of stored program computers, where a single store is used for both machine instructions and data. Its key components are as follows.

The key approach to building a general-purpose device according to von Neumann was in its ability to store not only its data and the intermediate results of computation, but also to store the instructions or commands for the computation. The computer instructions can be part of the hardware for specialised machines but for general-purpose machines, the computer instructions must be as changeable as the data that are acted upon by the instructions. *His insight was to recognise that both the machine instructions and data could be stored in the same memory.*

The key advantage of the von Neumann architecture over the existing approach was that it was much simpler to reconfigure a computer to perform a different task. All that was required was to enter new machine instructions in computer memory rather than physically rewiring a machine as was required with ENIAC. The limitations of von Neumann architecture include that it is limited to sequential processing and not very suitable for parallel processing.

Table 1.1 Von Neumann architecture

Component	Description
Arithmetic Unit	The arithmetic unit can perform basic arithmetic operations
Control Unit	The program counter contains the address of the next instruction to be executed. This instruction is fetched from memory and executed. This is the basic fetch and execute cycle (Fig. 1.4) The control unit contains a built-in set of machine instructions
Input–Output Unit	The input and output unit allows the computer to interact with the outside world
Memory	The one-dimensional memory stores all program instructions and data. These are usually kept in different areas of memory The memory may be written to or read from: i.e. it is random access memory (RAM) The program instructions are binary values, and the control unit decodes the binary value to determine which instruction to execute

1.5   Hardware and Software

Hardware is the physical part of the machine. It is tangible and may be seen or touched, and includes punched cards, vacuum tubes, transistors and circuit boards, integrated circuits and microprocessors. The hardware of a personal computer includes a keyboard, network cards, a mouse, a DVD drive, hard disk drive, printers and scanners and so on.

Software is intangible and consists of a set of instructions that tells the computer what to do. It is an intellectual creation of a programmer (or a team of programmers).

Operating system software manages the computer hardware and resources, and acts as an intermediary between the application programs and the computer hardware.

Application software refers to software programs that provide functionality for users to exploit the power of the computer to perform useful tasks such as business applications including spreadsheets and accountancy packages, financial applications, editors, compilers for programming languages, computer games, social media and so on.

1.6   Review Questions

1. Explain the difference between analog and digital computers.
2. Explain the difference between hardware and software.
3. What is a microprocessor?

4. Explain the difference between vacuum tubes, transistors and integrated circuits.
5. Explain the von Neumann architecture.
6. What are the advantages and limitations of the von Neumann architecture?
7. Explain the difference between a fixed program machine and a stored program machine.

1.7 Summary

A computer is a programmable electronic device that can process, store and retrieve data. It processes data according to a set of instructions or program. All computers consist of two basic parts namely the hardware and software. The hardware is the physical part of the machine, whereas software is intangible and is the set of instructions that tells the computer what to do.

There are two distinct families of computing devices namely digital computers and the historical analog computer. These two types of computer operate on quite different principles. The earliest digital computers were built in the 1940s, and these were large machines consisting of thousands of vacuum tubes. However, their computational power was a fraction of what is available today.

A digital computer is a sequential device that generally operates on data one step at a time. The data are represented in binary format, and a single transistor in a digital computer can store only two states: i.e. on and off. Several transistors are required to store larger numbers.

The representation of data in an analog computer reflects the properties of the data that is being modelled. Data and numbers may be represented by physical quantities such as electric voltage, whereas a stream of binary digits represents the data in a digital computer.

Von Neumann architecture is the fundamental architecture used on digital computers, and a single store is used for both machine instructions and data. Its introduction made it much easier to reconfigure a computer to perform a different task. All that was required was to enter new machine instructions in computer memory rather than physically rewiring a machine as was required with ENIAC.

Chapter 2
Foundations of Computing

Key Topics

Leibniz
Binary Numbers
Step Reckoner
Babbage
Difference Engine
Analytic Engine
Lovelace
Boole
Shannon
Switching Circuits

2.1 Introduction

This chapter considers important foundational work done by Wilhelm Leibniz, Charles Babbage, George Boole, Lady Ada Lovelace and Claude Shannon. Leibniz was a seventeenth-century German mathematician, philosopher and inventor, and he is recognised (with Isaac Newton) as the inventor of Calculus. He developed the Step Reckoner calculating machine that could perform all four basic arithmetic operations (i.e. addition, subtraction, multiplication and division), and he also invented the binary number system (which is used extensively in the computer field).

Boole and Babbage are considered grandfathers of the computing field, with Babbage's Analytic Engine providing a vision of a mechanical computer, and Boole's logic providing the foundation for modern digital computers.

Babbage was a nineteenth-century scientist and inventor who did pioneering work on calculating machines. He invented the Difference Engine (a sophisticated calculator that could be used to produce mathematical tables), and he also designed the Analytic Engine (the world's first mechanical computer). The design of the Analytic Engine included a processor, memory and a way to input information and output results.

© Springer International Publishing AG, part of Springer Nature 2018
G. O'Regan, *World of Computing*,
https://doi.org/10.1007/978-3-319-75844-2_2

Lady Ada Lovelace was introduced into Babbage's ideas on the Analytic Engine at a dinner party. She was fascinated and predicted that such a machine could be used to compose music, produce graphics, as well as solving mathematical and scientific problems. She explained how the Analytic Engine could be programmed, and she wrote what is considered the first computer program.

Boole was a nineteenth-century English mathematician who made important contributions to mathematics, probability theory and logic. Boole's logic provides the foundation for digital computers.

Shannon was the first person to apply Boole's logic to switching theory, and he showed that Boole's logic could simplify the design of circuits and telephone routing switches. It provides the perfect mathematical model for switching theory and for the subsequent design of digital circuits and computers.

2.2 Step Reckoner Calculating Machine

Leibniz (Fig. 2.1) was a German philosopher, mathematician and inventor in the field of mechanical calculators. He developed the binary number system used in digital computers, and he invented the Calculus independently of Sir Isaac Newton. He became familiar with Pascal's calculating machine, the *Pascaline*, while in Paris in the early 1670s. He recognised its limitations as the machine could perform addition and subtraction operations only.

He designed and developed a calculating machine that could perform addition, subtraction, multiplication, division and the extraction of roots. He commenced work on the machine in 1672, and the machine was completed in 1694. It was the first calculator that could perform all four arithmetic operations, and Leibniz's machine was called the *Step Reckoner* (Fig. 2.2). It allowed the common arithmetic operations to be carried out mechanically.

The operating mechanism used in his calculating machine was based on a counting device called the stepped cylinder or '*Leibniz wheel*'. This mechanism allowed a gear to represent a single decimal digit from zero to nine in just one revolution, and this remained the dominant approach to the design of calculating machines for the next 200 years. It was essentially a counting device consisting of a set of wheels that were used in calculation. The Step Reckoner consisted of an accumulator which could hold 16 decimal digits and an 8-digit input section. The eight dials at the front of the machine set the operand number, which was then employed in the calculation.

The machine performed multiplication by repeated addition and division by repeated subtraction. The basic operation is to add or subtract the operand from the accumulator as many times as desired. The machine could add or subtract an 8-digit number to the 16-digit accumulator to form a 16-digit result. It could multiply two 8-digit numbers to give a 16-digit result, and it could divide a 16-bit number by an

Fig. 2.1 Wilhelm Gottfried Leibniz

Fig. 2.2 Replica of Step Reckoner at Technische Sammlungen Museum, Dresden

8-digit number. Addition and subtraction are performed in a single step, with the operating crank turned in the opposite direction for subtraction. The result is stored in the accumulator.

2.3   Binary Numbers

Arithmetic has traditionally been done using the decimal notation[1], and Leibniz was one of the first to recognise the potential of the binary number system. This system uses just two digits namely '0' and '1', with the number two represented by 10; the number four by 100; and so on. Leibniz described the binary system in *Explication de l'Arithmétique Binaire* (Leibniz 1703), which was published in 1703. A table of values for the first fifteen binary numbers is given in Table 2.1.

Leibniz's 1703 paper describes how binary numbers may be added, subtracted, multiplied and divided, and he was an advocate of their use. The key advantage of the use of binary notation is in digital computers, where a binary digit may be implemented by an *on/off* switch, with the digit 1 representing that the switch is on, and the digit 0 representing that the switch is off.

The use of binary arithmetic allows more complex mathematical operations to be performed by relay circuits, and Boole's Logic (described in a later section) is the perfect model for simplifying such circuits, and is the foundation underlying digital computing.

The binary number system (base 2) is a positional number system, which uses two binary digits 0 and 1, and an example binary number is 1001.01_2 which represents $1 \times 2^3 + 0 \times 2^2 + 0 \times 2^1 + 1 \times 2^0 + 0 \times 2^{-1} + 1 \times 2^{-2} = 1 \times 2^3 + 1 \times 2^0 + 1 \times 2^{-2} = 8 + 1 + 0.25 = 9.25$.

The decimal system (base 10) is more familiar for everyday use, and there are algorithms to convert numbers from decimal to binary and vice versa. For example, to convert the decimal number 25 to its binary representation 11001_2, we proceed as follows (Fig. 2.3).

The base 2 is written on the left and the number to be converted to binary is placed in the first column. At each stage in the conversion, the number in the first column is divided by 2 to form the quotient and remainder, which are then placed on the next row. For the first step, the quotient when 25 is divided by 2 is 12 and the remainder is 1. The process continues until the quotient is 0, and the binary representation result is then obtained by reading the second column from the bottom up. Thus, we see that the binary representation of 25 is 11001_2.

Similarly, there are algorithms to convert decimal fractions to binary representation (to a defined number of binary digits as the representation may not terminate),

---

[1]The sexagesimal (or base-60) system was employed by the Babylonians c. 2000 BC. Indian and Arabic mathematicians developed the decimal system between 800 and 900 AD.

Table 2.1 Binary number system

Binary	Dec.	Binary	Dec.	Binary	Dec.	Binary	Dec.
0000	0	0100	4	1000	8	1100	12
0001	1	0101	5	1001	9	1101	13
0010	2	0110	6	1010	10	1110	14
0011	3	0111	7	1011	11	1111	15

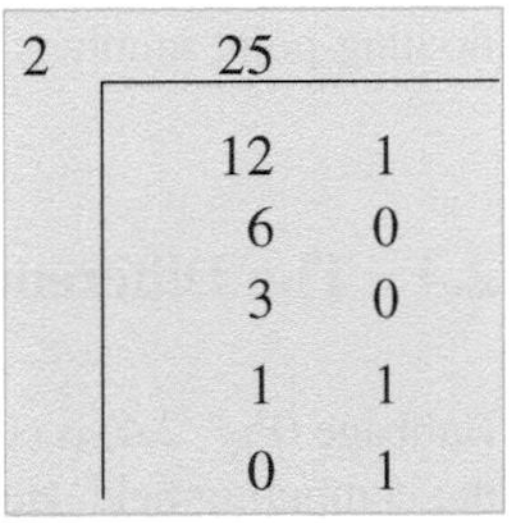

Fig. 2.3 Decimal to binary conversion

and the conversion of a number that contains an integer part and a fractional part involves converting each part separately and then combining them.

The octal (base 8) and hexadecimal (base 16) are often used in computing, as the bases 2, 8 and 16 are related bases and easy to convert between, as to convert between binary and octal involves grouping the bits into groups of three on either side of the point. Each set of three bits corresponds to one digit in the octal representation. Similarly, the conversion between binary and hexadecimal involves grouping into sets of four digits on either side of the point. The conversion from octal to binary or hexadecimal to binary is equally simple, and involves replacing the octal (or hexadecimal) digit with the 3-bit (or 4-bit) binary representation.

Numbers are represented in a digital computer as sequences of bits of fixed length (e.g. 16 bits, 32 bits). There is a difference in the way in which integers and real numbers are represented, with the representation of real numbers being more complicated.

An integer number is represented by a sequence (usually 2 or 4) bytes where each byte is eight bits. For example, a 2-byte integer has 16 bits with the first bit used as the sign bit (the sign is 1 for negative numbers and 0 for positive integers), and the remaining 15 bits represent the number. This means that two bytes may be used to represent all integer numbers between −32,768 and 32,767. A positive number is represented by the normal binary representation discussed earlier, whereas a negative number is represented using 2's complement of the original number (i.e. 0 changes to 1 and 1 changes to 0 and the sign bit is 1). All the standard arithmetic operations may then be carried out (using modulo 2 arithmetic).

The representation of floating-point real numbers is more complicated, and a real number is represented to a fixed number of significant digits (the significand) and scaled using an exponent in some base (usually 2). That is, the number is represented (approximated as):

$$\text{significand} \times \text{base}^{\text{exponent}}$$

The significand (also called mantissa) and exponent have a sign bit. For example, in simple floating-point representation (4 bytes), the mantissa is generally 24 bits and the exponent 8 bits, whereas for double precision (8 bytes) the mantissa is generally 53 bits and the exponent 11 bits. There is an IEEE standard for floating-point numbers (IEEE 754).

2.4 The Difference Engine

Babbage (Fig. 2.4) is considered (along with Boole) to be one of the grandfathers of the computing field. He made contributions to several areas including mathematics, statistics, astronomy, calculating machines, philosophy, railways and lighthouses. He founded the British Statistical Society and the Royal Astronomical Society.

Babbage was interested in accurate mathematical tables for scientific work. However, there was a high error rate in the existing tables due to human error introduced during calculation. He became interested in finding a mechanical method to perform calculation to eliminate the errors introduced by humans. He planned to develop a more advanced machine than the Pascaline or the Step Reckoner, and his goal was to develop a machine that could compute polynomial functions.

He designed the Difference Engine (No. 1) in 1821 for the production of mathematical tables. This was essentially a mechanical calculator (analogous to modern electronic calculators), and it was designed to compute polynomial functions of degree 4. It could also compute logarithmic and trigonometric functions such as sine or cosine (as these may be approximated by polynomials).[2]

The accurate approximation of trigonometric, exponential and logarithmic functions by polynomials depends on the degree of the polynomials, the number of decimal digits that it is being approximated to, and on the error function. A higher degree polynomial is generally able to approximate the function more accurately.

Babbage produced prototypes for parts of the Difference Engine, but he never actually completed the machine. The Swedish engineers, Georg and Edvard

---

[2]The power series expansion of the sine function is given by $\text{Sin}(x) = x - x^3/3! + x^5/5! - x^7/7! + \ldots$ The power series expansion for the cosine function is given by $\text{Cos}(x) = 1 - x^2/2! + x^4/4! - x^6/6! + \ldots$ Functions may be approximated by interpolation and the approximation of a function by a polynomial of degree n requires $n + 1$ points on the curve for the interpolation. That is, the curve formed by the polynomial of degree n that passes through the $n + 1$ points of the function to be approximated is an approximation to the function. The error function also needs to be considered.

Fig. 2.4 Charles Babbage

Scheutz, built the first working Difference Engine (based on Babbage's design) in 1853 with funding from the Swedish government. Their machine could compute polynomials of degree 4 on 15-digit numbers, and the 3rd Scheutz Difference Engine is on display at the Science Museum in London.

It was the first machine to compute and print mathematical tables mechanically. The machine was accurate, and it showed the potential of mechanical machines as a tool for scientists and engineers.

The machine is unable to perform multiplication or division directly. Once the initial value of the polynomial and its derivative are calculated for some value of x, the Difference Engine may calculate any number of nearby values using the numerical method of finite differences. This method replaces computational intensive tasks involving multiplication or division, by an equivalent computation that just involves addition or subtraction.

The British government cancelled Babbage's project in 1842. He designed an improved Difference Engine No. 2 (Fig. 2.5) in 1849. It could operate on seventh-order differences (i.e. polynomials of order 7) and 31-digit numbers. The machine consisted of eight columns with each column consisting of 31 wheels. However, it was over 150 years later before it was built (in 1991) to mark the two hundredth anniversary of his birth. The Science Museum in London also built the printer that Babbage designed, and both the machine and the printer worked correctly according to Babbage's design (after a little debugging).

2.5   The Analytic Engine—Vision of a Computer

The Difference Engine was designed to produce mathematical tables, but it required human intervention to perform the calculations. Babbage recognised its limitations, and he proposed a revolutionary solution by outlining his vision of a mechanical computer. His plan was to construct a new machine that would be capable of

Fig. 2.5 Difference Engine No. 2. Photo Public Domain

Table 2.2 Analytic Engine

Part	Function
Store	This contains the variables to be operated upon as well as all those quantities, which have arisen from the result of intermediate operations
Mill	The mill is essentially the processor of the machine into which the quantities about to be operated upon are brought

executing all tasks that may be expressed in algebraic notation. His vision of such a computer (Analytic Engine) consisted of two parts (Table 2.2).

Babbage intended that the operation of the Analytic Engine would be analogous to the operation of the *Jacquard loom*.[3] The latter is capable of weaving (i.e. executing on the loom) a design pattern that has been prepared by a team of skilled artists. The design pattern is represented by a set of cards with punched holes, where each card represents a row in the design. The cards are then ordered; placed in the loom; and the loom produces the exact pattern.

The use of the punched cards in the Analytic Engine allowed the formulae to be manipulated in a manner dictated by the programmer. The cards commanded the

---

[3]The Jacquard loom was invented by Joseph Jacquard in 1801. It is a mechanical loom which used the holes in punch cards to control the weaving of patterns in a fabric. The use of punched cards allowed complex designs to be woven from the pattern defined on the punched cards. Each punched card corresponds to one row of the design and the cards were appropriately ordered. It was very easy to change the pattern of the fabric being weaved on the loom, as this simply involved changing cards.

Analytic Engine to perform various operations and to return a result. Babbage distinguished between two types of punched cards:

– *Operation Cards*
– *Variable Cards.*

Operation cards are used to define the operations to be performed, whereas the variable cards define the variables or data that the operations are performed upon. His planned use of punched cards to store programs in the Analytic Engine is similar to the idea of a stored computer program in von Neumann architecture. However, Babbage's idea of using punched cards to represent machine instructions and data was over 100 years before digital computers. *Babbage's Analytic Engine is therefore an important milestone in the history of computing.*

Babbage intended that the program be stored on read-only memory using punch cards, and that the input and output would be carried out using punch cards. He intended that the machine would be able to store numbers and intermediate results in memory that could then be processed. There would be several punch card readers in the machine for programs and data. He envisioned that the machine would be able to perform conditional jumps as well as parallel processing where several calculations could be performed at once.

The Analytic Engine was designed in 1834 as the world's first mechanical computer (Lovelace 1842). It included a processor, memory, and a way to input information and output results. However, the machine was never built, as Babbage was unable to secure funding from the British Government.

2.5.1   Applications of Analytic Engine

Lady Augusta Ada Lovelace (nee Byron)[4] (Fig. 2.6) was a mathematician who collaborated with Babbage on applications for the Analytic Engine. She is considered the world's first programmer, and the Ada programming language is named in her honour.

She was introduced to Babbage at a dinner party in 1833, and she visited Babbage's studio in London, where the prototype Difference Engine was on display. She recognised the beauty of its invention, and she was fascinated by the idea of the Analytic Engine. She communicated regularly with Babbage with ideas on its applications.

Lovelace produced an annotated translation of Menabrea's '*Notions sur la machine analytique de Charles Babbage*' (Lovelace 1842). She added copious notes to the translation,[5] which were about three times the length of the original

---

[4]Lady Ada Lovelace was the daughter of the poet, Lord Byron.
[5]There is some controversy as to whether this was entirely her own work or a joint effort by Lovelace and Babbage.

Fig. 2.6 Lady Ada Lovelace

memoir, and considered many of the difficult and abstract questions connected with the subject. These notes are regarded as a description of a computer and software.

She explained in the notes how the Analytic Engine could be programmed, and wrote what is considered to be the first computer program. This program detailed a plan be written for how the engine would calculate Bernoulli numbers. Lady Ada Lovelace is therefore considered to be the first computer programmer, and Babbage called her the 'enchantress of numbers'.

She saw the potential of the Analytic Engine to fields other than mathematics. She predicted that the machine could be used to compose music, produce graphics, as well as solving mathematical and scientific problems. She speculated that the machine might act on other things apart from numbers, and be able to manipulate symbols according to rules. In this way, a number could represent an entity other than a quantity.

2.6   Boole's Symbolic Logic

George Boole (Fig. 2.7) was born in Lincoln, England in 1815. His father (a cobbler who was interested in mathematics and optical instruments) taught him mathematics, and showed him how to make optical instruments. Boole inherited his father's interest in knowledge, and he was self-taught in mathematics and Greek. He taught at various schools near Lincoln, and he developed his mathematical knowledge by working his way through Newton's Principia, as well as applying himself to the work of mathematicians such as Laplace and Lagrange.

Fig. 2.7 George Boole

He published regular papers from his early 20s, and these included contributions to probability theory, differential equations and finite differences. He developed Boolean algebra, which is the foundation for modern computing, and he is considered (along with Babbage) to be one of the grandfathers of computing. His work was theoretical, and he never actually built a computer or calculating machine. However, Boole's symbolic logic was the perfect mathematical model for switching theory and for the design of digital circuits.

Boole became interested in formulating a Calculus of reasoning, and he published 'Mathematical Analysis of Logic' in 1847 (Boole 1848). This work developed novel ideas on a logical method, and he argued that logic should be considered as a separate branch of mathematics, rather than as a part of philosophy. He argued that there are mathematical laws to express the operation of reasoning in the human mind, and he showed how Aristotle's syllogistic logic could be reduced to a set of algebraic equations. He corresponded regularly on logic with Augustus De Morgan.[6]

His paper on logic introduced two quantities '0' and '1'. He used the quantity 1 to represent the universe of thinkable objects (i.e. the universal set), and the quantity 0 represents the absence of any objects (i.e. the empty set). He then employed symbols such as x, y, z, etc., to represent collections or classes of objects given by the meaning attached to adjectives and nouns. Next, he introduced three operators (+, −, and ×) that combined classes of objects.

The expression xy (i.e. x multiplied by y or x × y) combines the two classes x, y to form the new class xy (i.e. the class whose objects satisfy the two meanings represented by class x and class y). Similarly, the expression x + y combines the

---

[6]De-Morgan was a nineteenth-century British mathematician based at University College London. De-Morgan's laws in Set Theory and Logic state that: $(A \cup B)^c = A^c \cap B^c$ and $\neg (A \vee B) \equiv \neg A \wedge \neg B$.

two classes x, y to form the new class x + y (that satisfies either the meaning represented by class x or class y). The expression x − y combines the two classes x, y to form the new class x − y. This represents the class (that satisfies the meaning represented by class x but not class y. The expression (1 − x) represents objects that do not have the attribute that represents class x.

Thus, if x = black and y = sheep, then xy represents the class of black sheep. Similarly, (1 − x) would represent the class obtained by the operation of selecting all things in the world except black things; x (1 − y) represents the class of all things that are black but not sheep; and (1 − x) (1 − y) would give us all things that are neither sheep nor black.

He showed that these symbols obeyed a rich collection of algebraic laws and could be added, multiplied, etc., in a manner that is like real numbers. These symbols may be used to reduce propositions to equations, and algebraic rules may be employed to solve the equations. The rules include the following:

1.	$x + 0 = x$	(Additive Identity)
2.	$x + (y + z) = (x + y) + z$	(Associative)
3.	$x + y = y + x$	(Commutative)
4.	$x + (1 − x) = 1$	
5.	$x\,1 = x$	(Multiplicative Identity)
6.	$x\,0 = 0$	
7.	$x + 1 = 1$	
8.	$xy = yx$	(Commutative)
9.	$x(yz) = (xy)z$	(Associative)
10.	$x(y + z) = xy + xz$	(Distributive)
11.	$x(y − z) = xy − xz$	(Distributive)
12.	$x^2 = x$	(Idempotent)

These operations are similar to the modern laws of set theory with the set union operation represented by '+', and the set intersection operation is represented by multiplication. The universal set is represented by '1' and the empty by '0'. The associative and distributive laws hold. Finally, the set complement operation is given by (1 − x).

He applied the symbols to encode Aristotle's Syllogistic Logic, and he showed how the syllogisms could be reduced to equations. This allowed conclusions to be derived from premises by eliminating the middle term in the syllogism. He refined his ideas on logic further in 'An Investigation of the Laws of Thought' which was published in 1854 (Boole 1958). This book aimed to identify the fundamental laws underlying reasoning in the human mind, and to give expression to these laws in the symbolic language of a Calculus.

He considered the equation $x^2 = x$ to be a fundamental law of thought. It allows the principle of contradiction to be expressed (i.e. for an entity to possess an attribute and at the same time not to possess it):

$$x^2 = x$$
$$\Rightarrow x - x^2 = 0$$
$$\Rightarrow x(1-x) = 0$$

For example, if x represents the class of horses then $(1 - x)$ represents the class of 'not-horses'. The product of two classes represents a class whose members are common to both classes. Hence, x $(1 - x)$ represents the class whose members are at once both horses and 'not-horses', and the equation x $(1 - x) = 0$ expresses that fact that there is no such class. That is, it is the empty set.

Boole contributed to other areas in mathematics including differential equations, finite differences[7] and to the development of probability theory. Des McHale has written an interesting biography of Boole (McHale 1985). Boole's logic appeared to have no practical use, but this changed with Claude Shannon's 1937 Master's Thesis, which showed its applicability to switching theory and to the design of digital circuits.

2.6.1 Switching Circuits and Boolean Algebra

Claude Shannon's Master's Thesis showed that Boole's algebra provided the perfect mathematical model for switching theory and for the design of digital circuits. It may be employed to optimise the design of systems of electromechanical relays, and circuits with relays solve Boolean algebra problems. The use of the properties of electrical switches to process logic is the basic concept that underlies all modern electronic digital computers. Digital computers use the binary digits 0 and 1, and Boolean logical operations may be implemented by electronic AND, OR and NOT gates. More complex circuits (e.g. arithmetic) may be designed from these fundamental building blocks.

Modern electronic computers use billions of transistors that act as switches and can change state rapidly. The use of switches to represent binary values is the foundation of modern computing. A high voltage represents the binary value 1 with low voltage representing the binary value 0. A silicon chip may contain billions of tiny electronic switches arranged into logical gates. The basic logic gates are AND, OR and NOT. These gates may be combined in various ways to allow the computer to perform more complex tasks such as binary arithmetic. Each gate has binary value inputs and outputs.

The example in Fig. 2.8 is that of an 'AND' gate which produces the binary value 1 as output only if both inputs are 1. Otherwise, the result will be the binary value 0. Figure 2.9 shows an 'OR' gate which produces the binary value 1 as output if any of its inputs is 1. Otherwise, it will produce the binary value 0.

---

[7]Finite Differences are a numerical method used in solving differential equations.

Fig. 2.8 Binary AND Operation

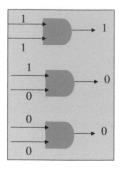

Fig. 2.9 Binary OR Operation

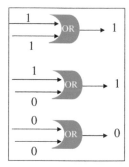

Fig. 2.10 NOT Operation

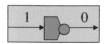

Fig. 2.11 Half-Adder

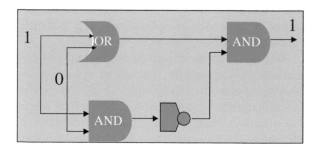

Finally, a NOT gate (Fig. 2.10) accepts only a single input which it inverts. That is, if the input is '1' the value '0' is produced and vice versa.

The logic gates may be combined to form more complex circuits. The example in Fig. 2.11 is that of a half-adder of 1 + 0. The inputs to the top OR gate are 1 and 0 which yields the result of 1. The inputs to the bottom AND gate are 1 and 0 which yields the result 0, which is then inverted through the NOT gate to yield binary 1.

Finally, the last AND gate receives two 1's as input and the binary value 1 is the result of the addition. The half-adder computes the addition of two arbitrary binary digits, but it does not calculate the carry. It may be extended to a full adder that provides a carry for addition.

2.7 Application of Symbolic Logic to Digital Computing

Claude Shannon (Fig. 2.12) was the first person[8] to see the applicability of Boole's algebra to simplify the design of circuits and telephone routing switches. He showed that Boole's symbolic logic was the perfect mathematical model for switching theory and for the subsequent design of digital circuits and computers.

His influential Master's Thesis is a key milestone in computing, and it shows how to lay out circuits according to Boolean principles. It provides the theoretical foundation of switching circuits, and his insight of using the properties of electrical switches to do Boolean logic is the basic concept that underlies all electronic digital computers.

Shannon realised that you could combine switches in circuits in such a manner as to carry out symbolic logic operations. This allowed binary arithmetic and more complex mathematical operations to be performed by relay circuits. He designed a circuit, which could add binary numbers, and he later designed circuits that could make comparisons and thus capable of performing a conditional statement. This was the birth of digital logic and the digital computing age.

Vannevar Bush (O'Regan 2013) was Shannon's supervisor at MIT, and Shannon's initial work was to improve Bush's mechanical computing device known as the Differential Analyser. This machine had a complicated control circuit that was composed of one hundred switches that could be automatically opened and closed by an electromagnet. Shannon's insight was his realisation that an electronic circuit is similar to Boolean algebra, and he showed how Boolean algebra could be employed to optimise the design of systems of electromechanical relays used in the analog computer. He also realised that circuits with relays could solve Boolean algebra problems.

His Master's thesis 'A Symbolic Analysis of Relay and Switching Circuits' (Shannon 1937) showed that the binary digits (i.e. 0 and 1) can be represented by electrical switches. This allowed binary arithmetic and more complex mathematical operations to be performed by relay circuits, and provided electronics engineers with the mathematical tool that they needed to design digital electronic circuits and provided the foundation for the field.

---

[8]Victor Shestakov at Moscow State University also proposed a theory of electric switches based on Boolean algebra (published in Russian in 1941 whereas Shannon's were published in 1937).

Fig. 2.12 Claude Shannon

The design of circuits and telephone routing switches could be simplified with Boole's symbolic algebra. Shannon showed how to lay out circuits according to Boolean principles, and his Master's thesis became the foundation for the practical design of digital circuits. These circuits are fundamental to the operation of modern computers and telecommunication systems, and his insight of using the properties of electrical switches to do Boolean logic is the basic concept that underlies all electronic digital computers.

2.8   Review Questions

1. Explain the significance of binary numbers in the computing field.
2. Explain the importance of Shannon's Master Thesis.
3. Explain the significance of the Analytic Engine.
4. Explain why Ada Lovelace is considered the world's first programmer.
5. Explain the significance of Boole to the computing field.
6. Explain the significance of Babbage to the computing field.
7. Explain the significance of Leibniz to the computing field.

2.9 Summary

This chapter considered the foundational work done by Leibniz, Babbage, Boole, Ada Lovelace and Shannon. Leibniz developed a calculating machine (the Step Reckoner) that could perform the four basic arithmetic operations. He also invented the binary number system, which is used extensively in the computer field.

Babbage did pioneering work on calculating machines. He designed the Difference Engine (a sophisticated calculator that could be used to produce mathematical tables), and he also designed the Analytic Engine (the world's first mechanical computer).

Lady Ada Lovelace was introduced to Babbage's ideas on the Analytic Engine, and she predicted that such a machine could be used to compose music, produce graphics, as well as solving mathematical and scientific problems.

Boole was a nineteenth-century English mathematician who made important contributions to mathematics, and his symbolic logic provides the foundation for digital computers.

Shannon was a twentieth-century American mathematician and engineer, and he showed that Boole's symbolic logic provided the perfect mathematical model for switching theory, and for the subsequent design of digital circuits and computer.

Chapter 3
A Concise History of Computing

Key Topics

Harvard Mark I
ABC computer
ENIAC and EDVAC
Colossus
Zuse's Machines
Manchester Mark I
SAGE
System 360
Transistor
Integrated Circuits
Microprocessor
Home Computers
IBM Personal Computer

3.1 Introduction

The objective of this chapter is to give a brief account of the computing field from its beginnings up to the development of the IBM personal computer in the early 1980s. We discuss some of the early computers developed in the United States, Britain and Germany in the 1940s, where the objective was to provide faster methods of computation to solve practical (often military) problems. These large bulky machines were mainly based on vacuum tube technology.

We discuss the invention of the transistor in the late 1940s, and this led to smaller more powerful machines. We proceed to the development of integrated circuits in the late 1950s, and the later development of minicomputers and more powerful mainframes. We discuss the introduction of the IBM System/360 in the 1960s, and its impact on the computer sector.

We then proceed to the invention of the microprocessor and the subsequent development of home computers. Finally, we discuss the introduction of the IBM personal computer in the early 1980s, and the introduction of the Apple Macintosh in the mid-1980s.

© Springer International Publishing AG, part of Springer Nature 2018
G. O'Regan, *World of Computing*,
https://doi.org/10.1007/978-3-319-75844-2_3

3.2   Early Digital Computers

The Second World War motivated researchers to investigate faster ways to perform calculation to solve practical problems. This led research into the development of digital computers to determine if they could assist in speeding up computation, and the early computers were large bulky machines consisting of several thousand vacuum tubes. A computer often took up the space of a large room, and it was slow and unreliable.

The early computers considered include the Harvard Mark I designed and developed by Howard Aiken and IBM. John Atanasoff and Clifford Berry designed and developed the Atanasoff–Berry (ABC) computer, and it was designed to solve a set of linear equations using Gaussian elimination. John Mauchly and Presper Eckert designed the ENIAC and EDVAC computers, which was a fixed program computer that needed to be physically rewired to solve different problems. Its successor, the EDVAC computer, implemented the concept of a stored program,[1] and did not need to be physically rewired to solve a new problem.

The team at Bletchley Park in England designed and developed the Colossus computer as part of their code-breaking work during the Second World War. This machine allowed them to crack the German Lorenz codes, and to provide important military intelligence for the D-Day landings in 1944.

Konrad Zuse designed and developed the Z1, Z2 and Z3 machines in Germany. The Z3 was operational in 1941, and it was the world's first programmable computer. Zuse's machines were based on relays rather than vacuum tube technology.

3.2.1   Harvard Mark 1

Howard Aiken (Fig. 3.1) made several important contributions to the early computing field. He showed that a large calculating machine could be built that would provide speedy solutions to mathematical problems.

He designed and constructed an electromechanical machine to perform mathematical operations quickly and efficiently, and the machine could handle positive and negative numbers, scientific functions such as logarithms and was able to work with minimal human intervention.

He discussed the idea with colleagues and IBM, and he was successful in obtaining IBM funding to build the machine. The machine was built at the IBM laboratories at Endicott with several IBM engineers involved in its construction. The construction took seven years, and it was completed in late 1943.

---

[1]This meant that the program instructions were stored in memory and all that was required to carry out a new task was to load a new program into memory.

Fig. 3.1 Howard Aiken

The machine became known as the Harvard Mark I [also known as the IBM *Automatic Sequence Controlled Calculator* (ASCC)]. Aiken was influenced by Babbage's ideas on the design of the Analytic Engine.

IBM presented the machine to Harvard University in 1944, and the ASCC was essentially an electromechanical calculator that could perform large computations automatically. It could perform addition, subtraction, multiplication and division, and it could refer to previous results.

The Harvard Mark I (Fig. 3.2) was designed to assist in the numerical computation of differential equations. It was 50 feet long, 8 feet high and weighed 5 tons. It performed additions in less than a second, multiplications in 6 s, and division in

The Harvard Mark I

Fig. 3.2 Harvard Mark I (IBM ASCC). Courtesy of International Business Machines Corporation, © International Business Machines Corporation

about 12 s. It used electromechanical relays to perform the calculations, and it could execute long computations automatically.

It was constructed out of switches, relays, rotating shafts and clutches, and it used 500 miles of wiring and over 750,000 components. It was the industry's largest electromechanical calculator, and it had 60 sets of 24 switches for manual data entry. It could store 72 numbers, each 23 decimal digits long. The instructions were read on paper tape, and punched cards were used to input the data, and the results were either on punched cards or an electric typewriter.

The US Navy used the Harvard Mark I for ballistic calculations, and the machine remained in use until 1959. It cost approximately half a million dollars, but it was never mass-produced by IBM. It differed from most of the early digital computers as it used relays instead of vacuum tubes.

The announcement of the Harvard Mark 1 led to tension between Aiken and IBM, as Aiken announced himself as the sole inventor without acknowledging the important role played by IBM.

3.2.2   Atanasoff–Berry Computer

John Atanasoff (Fig. 3.3) was born in New York in 1903. He studied electrical engineering at the University of Florida and did a Master's in Mathematics at Iowa State College. He earned a PhD in theoretical physics from the University of Wisconsin in 1930 and became an assistant professor at Iowa State College, where he taught mathematics and physics.

He became interested in developing faster methods of computation while doing his Ph.D. research. He did some work on an analog calculator in 1936, but he concluded that analog devices were too restrictive and unable to give the desired accuracy. His goal was to mechanise calculation to enable accurate computation to be carried out faster.

The existing computing devices were mechanical, electromechanical or analog. Atanasoff developed the concept of digital machine in the late 1930s, and he believed that his proposed machine would be faster and more accurate than the existing analog machines. He published his design of a machine to solve linear equations using his own version of Gaussian elimination in 1939 and used his research grant of $650 to build the Atanasoff–Berry Computer (ABC), with the assistance of his graduate student, Clifford Berry, from 1939 to 1942.

The ABC (Fig. 3.4) was approximately the size of a large desk and had approximately 270 vacuum tubes. Two hundred and ten tubes controlled the arithmetic unit, 30 tubes controlled the card reader and card punch, and the remaining tubes helped maintain charges in the condensers. It employed rotating drum memory, with each of the two drum memory units able to hold thirty 50-bit numbers.

Fig. 3.3 John Atanasoff with components of ABC

Fig. 3.4 Replica of ABC computer. Courtesy of Iowa State University

It was designed for a specific purpose (i.e. solving linear equations) rather than as a general-purpose computer. The working prototype was one of the earliest electronic digital computers.[2] However, the ABC was slow, and it required constant operator monitoring.

It used binary mathematics and Boolean logic to solve simultaneous linear equations. It employed over 270 vacuum tubes for digital computation, but it had no central processing unit (CPU), and it was not programmable.

It weighed over 300 kg and it used 1.6 km of wiring. It used 50-bit numbers, and it could perform 30 additions or subtractions per second. The memory and arithmetic units could operate and store 60 such numbers at a time (60 * 50 = 3000 bits). The arithmetic logic unit was fully electronic, and it was implemented with vacuum tubes.

The input was in decimal format with standard IBM 80 column punch cards, and the output was in decimal format via a front panel display. A paper card reader was used as an intermediate storage device to store the results of operations that were too large to be handled entirely within electronic memory. The ABC pioneered important elements in modern computing including:

- Binary arithmetic and Boolean logic
- All calculations were performed using electronics rather than mechanical switches
- Computation and memory were separated.

The ABC was tested and operational by 1942, and its historical significance is that it demonstrated the feasibility of electronic computing. Several of its concepts were later used in the ENIAC computer developed by Mauchly and Eckert.

3.2.3   ENIAC

The Electronic Numerical Integrator and Computer (ENIAC) (Fig. 3.5) was one of the first large general-purpose digital computers. It was used to integrate ballistic equations, and to calculate the trajectories of naval shells. It was completed in 1946, and it remained in use until 1955.

It was a large bulky machine (100 feet long, 10 feet high, 3 feet deep and weighed about 30 tons). Its development commenced in 1943 at the University of Pennsylvania, and it was built for the US Army's Ballistics Research Laboratory The project team included Presper Eckert as chief engineer and John Mauchly as a consultant. ENIAC had over 18,000 vacuum tubes, and the machine generated a vast quantity of heat, as each vacuum tube generated heat like a light bulb. The machine used 150 kW of power and air conditioning was used to cool it.

---

[2]The ABC was ruled to be the first electronic digital computer in the Sperry Rand versus Honeywell patent case in 1973. However, Zuse's Z3 computer (completed in Germany in 1941) preceded it.

Fig. 3.5 Setting the switches on ENIAC's Function Tables. U.S. Army Photo

It employed decimal numerals and it could add five thousand numbers, do over three hundred and fifty 10-digit multiplications, or thirty-five 10-digit divisions in one second. It could be programmed to perform complex sequences of operations, and this included loops, branches and subroutines. However, the task of taking a problem and mapping it onto the machine was complex, and it usually took weeks to perform. The first step was to determine what the program was to do on paper; the second step was the process of manipulating the switches and cables to enter the program into ENIAC, and this usually took several days. The final step was verification and debugging, and this often involved single step execution of the machine.

There were problems initially with the reliability of ENIAC, as several vacuum tubes burned out most days (Fig. 3.6). This meant that the machine was often non-functional, as high-reliability vacuum tubes only became available in the late 1940s. However, most of these problems with the tubes occurred during the warm-up and cool-down periods, and it was therefore decided not to turn the machine off. This led to improvements in its reliability to the acceptable level of one tube every two days. The longest continuous period of operation without a failure was five days.

The very first program run on ENIAC took just 20 s, and the answer was manually verified to be correct after forty hours of work with a mechanical calculator. One of the earliest problems solved was related to the feasibility of the hydrogen bomb, and this program involved the input of 500,000 punch cards, and it ran for six weeks and gave an affirmative reply.

ENIAC was a *fixed-program* computer, and the machine had to be physically rewired to perform different tasks. It was clear that there was a need for an

Fig. 3.6 Replacing a valve on ENIAC. U.S. Army Photo

architecture that would allow a machine to perform different tasks without physical rewiring each time. This led to the concept of the *stored program*, which was implemented on EDVAC (the successor to ENIAC).

The idea of a stored program is that the program is stored in memory, and whenever there is a need to change the task that is to be computed, then all that is required is to place a new program in the memory of the computer, rather than rewiring the machine. EDVAC implemented the concept of a stored program in 1949, just after its implementation on the Manchester Baby prototype machine in England. The concept of a stored program and von Neumann architecture is detailed in von Neumann's report on EDVAC (von Neumann 1945).

ENIAC was preceded in development by Zuses's Z3 machine in Germany; the Atanasoff–Berry Computer (ABC) in the United States; and the Colossus computer developed in the UK. ENIAC was a major milestone in the history of computing.

3.2.4   EDVAC

The Electronic Discrete Variable Automatic Computer (EDVAC) (Fig 3.7) was the successor to the ENIAC computer. It was a stored program computer and it was proposed by Eckert and Mauchly in 1944, and design work commenced prior to the completion of ENIAC.

Fig. 3.7 The EDVAC
Computer. U.S. Army Photo

It was delivered to the Ballistics Research Laboratory in 1949, and it commenced operations in 1951. It remained in operations until 1961. It employed 6000 vacuum tubes and its power consumption was 56,000 W. It had 5.5 Kb of memory.

EDVAC was one of the earliest stored-program computers, and the program instructions were stored in memory, rather than rewiring the machine each time.

3.2.5   Bletchley Park and Colossus

Tommy Flowers (Fig. 3.8) was a British engineer who led the team that designed and built Colossus, which was one of the earliest electronic computers. The machine was designed to decode the top-level encrypted German military communication sent by German High Command to its commanders in the field. This provided British and American Intelligence with important information on German military plans around the D-Day invasion and later battles, and it helped to ensure the success of the Normandy landings and the ultimate defeat of Nazi Germany.

Flowers was born in East London in 1905, and he obtained a position with the telecommunications branch of the General Post Office in 1926. He moved to the research station at Dollis Hill in 1930, and he investigated the use of electronics for telephone exchanges. He was convinced at an early stage that an all-electronic system was possible.

He became involved with the code-breaking work taking place at Bletchley Park (located near Milton Keynes north-west of London) during the Second World War.

Fig. 3.8 Tommy Flowers

Alan Turing and others had cracked the German Enigma codes by building a machine known as the Bombe. This machine employed a crib to deduce the settings of the Enigma machine for that day. Turing introduced Flowers to Max Newman who was leading British efforts to break a German cipher generated by the Lorenz SZ42 machine.

Their existing approach to deciphering the Lorenz codes was with the Heath Robinson machine (a slow and unreliable machine). Flowers proposed an alternate solution involving the use of an electronic machine in 1943. This machine was called Colossus and it employed 1800 thermionic valves. The management at Bletchley Park was sceptical, but they encouraged him to continue with his work. Flowers and others at the Post Office Research Centre built the machine in 11 months, and its successor, the Mark 2 Colossus, contained 2400 valves. This machine commenced operations on June 1st, 1944. It was a large bulky machine and took up the space of a small room and weighed a ton.

It provided vital information for the Normandy landings, and it confirmed that Hitler had been successfully misled by Allied disinformation into believing that the Normandy landings were to be a diversionary tactic. Further, it confirmed that no additional German troops were to be moved there, and it played a key role in helping the British to monitor the German reaction to their deception tactics.

Flowers and others designed and built the original Colossus machine at the Post Office Research Station at Dollis Hill in London. The machine was used to find possible key combinations for the Lorenz machines rather than decrypting an intercepted message in its entirety. The Lorenz machine was based on the *Vernam cipher*.

Colossus compared two data streams to identify possible key settings for the Lorenz machine. The first data stream was the encrypted message, and it was read at high speed from a paper tape. The second stream was generated internally and was an electronic simulation of the Lorenz machine at various trial settings. If the match count for a setting was above a certain threshold, it would be sent as output to an electric typewriter.

The Lorenz codes were a more complex cipher than the Enigma codes, and they were used in the transmission of important messages between the German High Command in Berlin and the military commanders in the field. The Lorenz SZ 40/42 machine performed the encryption. The Bletchley Park codebreakers called the typewriter-coding machine '*Tunny*' and the coded messages '*Fish*'.

The Colossus Mark 1 machine was specifically designed for codebreaking rather than as a general-purpose computer. It was semi-programmable and helped in deciphering messages encrypted using the Lorenz machine. A prototype was available in 1943 and a working version was available in early 1944 at Bletchley Park. The Colossus Mark 2 (Fig. 3.9) was introduced just prior to the Normandy landings in June 1944.

The Colossus Mark 1 used 15 kW of power and it could process 5000 characters of paper tape per second. It enabled a large amount of mathematical work to be done in hours rather than in weeks. There were ten Colossi machines working at Bletchley Park by the end of the war. A replica of the Colossus was rebuilt by a team of volunteers led by Tony Sale from 1993 to 96, and it is at Bletchley Park museum.

The contribution of Bletchley Park to the cracking of the German Enigma and Lorenz codes, and to early computing remained clouded in secrecy until recent times. The museum at Bletchley Park provides insight into the important contribution made by this organisation to codebreaking and to early computing during the Second World War.

Fig. 3.9 Colossus.Mark 2. Photo Courtesy of UK Government

3.2.6   Zuse's Machines

Konrad Zuse (Fig. 3.10) is considered *'the father of the computer'* in Germany, as he built the world's first programmable machine (the Z3) in 1941.

He was born in Berlin in 1910, and he studied civil engineering at the Technical University of Berlin. He commenced working for Henschel (an airline manufacturer) after his graduation in 1935. He resigned after one year with the intention of forming his own company to build automatic calculating machines, and he commenced work on what would become the Z1 machine in 1936. Zuse employed the binary system for the calculator and metallic shafts that could shift from position 0 to 1 and vice versa. The Z1 was operational by 1938.

He served in the German Army on the Eastern Front for six months in 1939 at the start of the Second World War. Henschel helped Zuse to obtain a deferment from the army, and he remained affiliated with Henschel for the duration of the war. He built the Z2 and Z3 machines during this period, and the Z3 was operational in 1941, and it was the world's first programmable computer.

He started his own company in 1941, and this was the first company founded with the sole purpose of developing computers. The Z4 was almost complete as the Red Army advanced on Berlin in 1945, and Zuse left Berlin for Bavaria with the Z4. His other machines were destroyed in the Allied bombing of Germany.

He designed the world's first high-level programming language (Plankalkül) between 1943 and 1945. He later restarted his company (Zuse KG), and he completed the Z4 in 1950. This was the first commercial computer, as it was completed ahead of the Ferranti Mark 1, Univac and LEO computers. Its first customer was the Technical University of Zurich.

Fig. 3.10   Konrad Zuse.
Courtesy of Horst Zuse,
Berlin

Zuse's results are all the more impressive given that he was working alone in Germany, and was unaware of developments in other countries. He independently implemented the principles of modern digital computers in isolation (O'Regan 2013).

3.2.7   Z1, Z2 and Z3 Machines

Zuse commenced work on his first machine called the Z1 in 1936, and the machine was operational by 1938. It was demonstrated to a small number of people who saw it rattle and compute the determinant of a three by three matrix. It was essentially a binary electrically-driven mechanical calculator with limited programmability. It could execute instructions read from the program punch cards, but the program itself was never loaded into the memory.

It employed the binary system and metallic shafts that could slide from position 0 to position 1 and vice versa. The machine was essentially a 22-bit floating-point value adder and subtracter. A decimal keyboard was used for input, and the output was decimal digits. The machine included some control logic, which allowed it to perform more complex operations such as multiplications and division. These operations were performed by repeated additions for multiplication and repeated subtractions for division. The multiplication took approximately 5 s. The computer memory contained 64 22-bit words. Each word of memory could be read from and written to by the program punch cards and the control unit. It had a clock speed of 1 Hz, and two floating-point registers of 22 bits each. The machine was unreliable, and there is a reconstruction of it in the Deutsches Technikmuseum in Berlin.

His Z2 machine aimed to improve on the Z1, and this mechanical and relay computer was created in 1939. It used a similar mechanical memory, but it replaced the arithmetic and control logic with 600 electrical relay circuits. It used 16-bit fixed-point arithmetic instead of the 22-bit used in the Z1. It had a 16-bit word size and the size of its memory was 64 words. It had a clock speed of 3 Hz.

The Z3 machine (Fig. 3.11) was the first functional tape-stored-program-controlled computer, and it was created in 1941. It used 2600 telephone relays, the binary number system, and it could perform floating-point arithmetic. It had a clock speed of 5 Hz, and multiplication and division took 3 s. The input to the machine was with a decimal keyboard, and the output was on lamps that could display decimal numbers. The word length was 22-bits, and the size of the memory was 64 words.

It used a punched film for storing the sequence of program instructions. It could convert decimal to binary and back again. It was the first digital computer since it predates the Atanasoff–Berry Computer by one year. It was proven to be Turing complete in 1998. There is a reconstruction of the Z3 computer in the Deutsches Museum in Munich.

Fig. 3.11 Zuse and the Reconstructed Z3. Courtesy of Horst Zuse, Berlin

3.2.8 University of Manchester

The Manchester Small-Scale Experimental Computer (better known by its nick-name 'Baby') was developed at the University of Manchester. It was the *first stored-program computer*, and it was designed and built at Manchester University in England by Frederic Williams, Tom Kilburn, Geoff Tootill and others.

The task to be computed is defined by the computer instructions that are placed in memory, and to change the task to be computed, all that is required is to load a different program into the computer memory. Kilburn wrote and executed the first stored program, and it was a short 17-line program written and executed in 1948.

The prototype 'Baby' (Fig. 3.12) demonstrated the feasibility and potential of a stored program computer. Its memory consisted of 32 32-bit words, and it took 1.2 ms to execute one instruction: i.e. 0.00083 MIPS (million instructions per second). Today's computers are rated at speeds of up to 1000 MIPS and more). The team in Manchester developed the machine further and in 1949, the Manchester Mark 1 was available.

3.2.9 Manchester Mark I

The Manchester Automatic Digital Computer (MADC), also known as the Manchester Mark 1, was developed at the University of Manchester. It was one of the earliest stored-program computers, and it was the successor to the Manchester 'Baby' computer. It was designed and built by Williams, Kilburn and others.

Fig. 3.12 Replica of the Manchester Baby. Courtesy of Tommy Thomas

Each word could hold one 40-bit number or two 20-bit instructions. The main memory consisted of two pages (i.e. two Williams tubes with each holding 32 × 40-bit words or 1280 bits). The secondary backup storage was a magnetic drum consisting of 32 pages (this was updated to 128 pages in the final specification). Each track consisted of two pages (2560 bits). One revolution of the drum took 30 ms, and this allowed the 2560 bits to be transferred to main memory.

The Manchester Mark I (Fig. 3.13) contained 4050 vacuum tubes, and it had a power consumption of 25,000 W. The standard instruction cycle was 1.8 ms but multiplication was much slower. The machine had 26 defined instructions, and the programs were entered into the machine in binary format, as assembly languages and assemblers were not yet available.

It had no operating system and its only systems software were some basic routines for input and output. Its peripheral devices included a teleprinter and a 5-hole paper tape reader and punch.

A display terminal used with the Manchester Mark 1 computer mirrored what was happening within the Williams Tube. A metal detector plate placed close to the surface of the tube detected changes in electrical charges. The metal plate obscured a clear view of the tube, but the technicians could monitor the tubes used with a video screen. Each dot on the screen represented a dot on the tube's surface, and the

Fig. 3.13 The Manchester Mark 1 Computer. Courtesy of the University of Manchester

dots on the tube's surface worked as capacitors that were either charged and bright or uncharged and dark. The information translated into binary code (0 for dark, 1 for, bright) became a way to program the computer.

The Manchester Mark I influenced later computer development such as Ferranti's Mark I general-purpose computer which was released in 1951, as well as early IBM computers such as the IBM 701.

3.3   Early Commercial Computers

This section considers a selection of computers developed during the 1950s, and it includes a selection of vacuum tube-based computers as well as transistor computers. One of the drivers for the design and development of more powerful computers was the perceived threat of the Soviet Union. This led to an arms race between the two superpowers, and it was clear that computing technology would play an important role in developing more sophisticated weapon and defence systems. The SAGE air defence system developed for the United States and Canada was an early example of the use of computer technology for the military.

The other key driver for the development of more powerful computers was to support business, universities and government. The machines developed during this period were mainly large proprietary mainframes designed for business, scientific and government use. They were expensive and this eventually led vendors such as IBM and DEC to introduce families of computers in the 1960s, where a customer could choose a small cheaper member of the family, and to upgrade over time to a larger computer as their needs evolved.

3.3.1 The SAGE System

The Semi-Automated Ground Environment (SAGE) was an automated system for tracking and intercepting enemy aircraft in North America. It was used by the North American Aerospace Defense Command (NORAD) from the late 1950s until the 1980s. The interception of enemy aircraft was extremely difficult prior to the invention of radar during the Second World War. Its introduction allowed fighter aircraft to be scrambled just in time to meet the enemy threat. The radar stations were ground-based, and they needed to communicate with and send interception instructions to fighter aircraft to deal with hostile aircraft.

However, after the war the speed of aircraft increased considerably, thereby reducing the time available to scramble fighter aircraft. This necessitated a more efficient and automatic way to transmit interception instructions to provide security to the United States. The SAGE system (Fig. 3.14) was designed to solve this problem, and it analysed the information that it received from the various radar stations around the country in real-time. *It then automated the transmission of interception messages to fighter aircraft.*

IBM and MIT played an important role in the design and development of SAGE. Some initial work on real-time computer systems had been done at Massachusetts Institute of Technology on a project for the United States Navy. This project was concerned with building an aircraft flight simulator computer for training bombing crews, and it led to the development of the Whirlwind digital computer. This computer was originally intended to be an analog machine, but instead, it became the Whirlwind digital computer. It was used for experimental development of military combat information systems.

Whirlwind was the first real-time computer, and George Valley and Jay Forrester wrote a proposal to employ Whirlwind for air defence. This led to the Cape Cod system, which demonstrated the feasibility of an air defence system covering New England. The design and development of SAGE commenced in 1953.

IBM was responsible for the design and manufacture of the AN/FSQ-7 vacuum tube computer used in SAGE. Its design was based on the Whirlwind II computer, which was intended to be the successor to Whirlwind. However, the Whirlwind II was never built, and the AN/FSQ-7 computer weighed 275 tons and included 500,000 lines of assembly code.

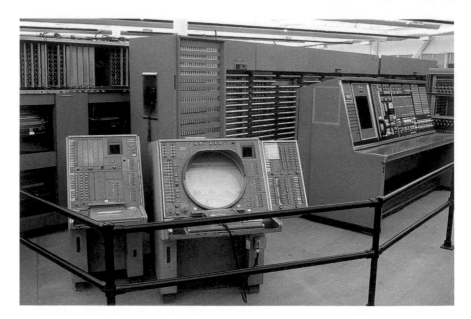

Fig. 3.14 SAGE. Photo Courtesy of Steve Jurvetson

The AN/FSQ holds the current world record for the largest computer ever built. It employed 55,000 vacuum tubes; covered an area over 18,000 square feet, and it used about three megawatts of power.

There were twenty-four SAGE Direction Centres and three SAGE Combat Centres located in the United States. Each SAGE site included two computers for redundancy, and long-distance telephone lines linked each centre. Burroughs provided the communications equipment to enable the centres to communicate with one another, and *this was one of the earliest computer networks*.

Each site was connected to multiple radar stations with tracking data transmitted by modem over a standard telephone wire. The SAGE computers then collected the tracking data for display on a cathode ray tube (CRT). The console operators at the centre could select any of the targets on the display to obtain information on the tracking data. This enabled aircraft to be tracked and identified, and the electronic information was presented to operators on a display device.

The engineering effort in the SAGE project was immense and the total cost is believed to have been several billion US dollars. It was a massive construction project, which involved erecting buildings and building power lines, and communication links between the various centres and radar stations.

SAGE influenced the design and development of the Federal Aviation Authority (FAA) automated air traffic control system.

3.3.2 *Invention of the Transistor*

The early computers were large bulky machines taking up the size of a large room. They contained thousands of vacuum tubes,[3] and these tubes consumed large amounts of power and generated a vast quantity of heat. This led to problems with the reliability of the early computers, as several tubes burned out each day. This meant that machines were often nonfunctional for parts of the day until the defective tube was identified and replaced (see Fig. 3.6).

Therefore, there was a need to find a better solution, and Shockley (Fig. 3.15) set up the solid physics research group at Bell Labs after the Second World War. His goal was to find a solid-state alternative to the existing glass-based vacuum tubes. His solid physics research team included John Bardeen and Walter Brattain, who would later share the 1956 Nobel Prize in Physics with him for their invention of the transistor (Fig. 1.2).

Their early research was unsuccessful, but by late 1947 Bardeen and Brattain succeeded in creating a point contact transistor independently of Shockley, who was working on a junction-based transistor. Shockley believed that the point contact transistor would not be commercially viable, and his junction point transistor was announced in mid-1951. The junction point transistor soon eclipsed the point-contact transistor, and it became dominant in the marketplace.

Shockley published a book on semiconductors in 1950 (Shockley 1950), and he resigned from Bell Labs in 1955. He formed Shockley Laboratory for Semiconductors (part of Beckman Instruments) at Mountain View in California. This company played an important role in the development of transistors and semiconductors, and several of its staff later formed semiconductor companies in the Silicon Valley area.

Shockley was the director of the company but his management style alienated several of his employees. This led to the resignation of eight key researchers in 1957 following his decision not to continue research into silicon-based semiconductors. This gang of eight went on to form Fairchild Semiconductors and other companies in the Silicon Valley area in the following years (O'Regan 2016a).

3.3.3 *Early Transistor Computers*

The University of Manchester Experimental Transistor Computer was one of the first transistor-based computers.[4] The prototype machine used 92 point-contact transistors and had a 48-bit word size, whereas the full-scale version used 200

---

[3]ENIAC contained over 18,000 vacuum tubes and the AN/FSQ-7 computer used in SAGE contained 55,000 vacuum tubes.

[4]It was not a fully transistorised computer in that it employed a small number of vacuum tubes in its clock generator.

Fig. 3.15 William Shockley.
Courtesy Chuck Painter,
Stanford News Service

point-contact transistors. There were serious problems with the reliability of the point-contact transistors, and Metropolitan-Vickers (a Manchester company) adapted the design and changed the circuits to use the more reliable junction-based transistors. This led to the development of the full-scale version called the Metrovick 950 in 1956.

The TRADIC computer was designed and developed by Bell Labs in early 1954, and it included both transistors and vacuum tubes. The Harwell CADET was an early fully transistorised machine when it appeared in early 1955. The IBM 608 was the first IBM product to use transistor circuits, and the fully transistorised calculator was introduced in late 1957. The Burroughs SM-65-Atlas ICBM was an early transistorised computer, which appeared in 1957.

The IBM 7090 was one of the earliest commercial computers with transistor logic, and it was introduced in 1958. It was designed for large-scale scientific applications, and it was over thirteen times faster than the older vacuum tube IBM 701. It used 36-bit words, had an address space of 32,768 words, and could perform 229,000 calculations per second. It was used by the U.S. Air Force to provide an early warning system for missiles, and by NASA to control space flights.

3.4   Integrated Circuits

The invention of the transistor was a revolution in computing, and it led to smaller, faster and more reliable computers. However, it was still a challenge for engineers to design complex circuits, as they had to wire hundreds (thousands) of separate components together.

It is essential when building a circuit that all the connections are intact, otherwise the electric current will be stopped on its way through the circuit, and the circuit will fail. Prior to the invention of the integrated circuit, engineers had to construct circuits by hand, which involved soldering each component in place and connecting them with wires. However, the manual assembly of the large number of components required in a computer often resulted in faulty connections, and advanced computers required so many connections that they were almost impossible to build. Clearly, there was a need for a better solution.

The invention of the integrated circuit allowed many transistors to be combined on a single chip, and it was another revolution in computing. The integrated circuit placed the previously separated transistors, resistors, capacitors and wiring circuitry onto a single chip made of silicon or germanium. The integrated circuit shrunk the size and cost of making electronics, and it had a major influence on the design of later computers and electronics. It led to faster and more powerful computers.

3.4.1 Invention of Integrated Circuit

Transistors were tiny in comparison to vacuum tubes, they consumed very little power and they were more reliable. This stimulated engineers to design ever more complex electronic circuits containing hundreds or thousands of discrete components such as transistors, diodes, rectifiers and capacitors.

This meant that engineers faced problems in increasing the performance of their designs as the number of components in the design increased. Each component needed to be wired to many other components, and the wiring and soldering were done manually. Clearly, more components would be required to improve performance, and therefore, it seemed that future designs would consist almost entirely of wiring.

The hand soldering of thousands of components to thousands of bits of wire was expensive and time-consuming, and it was also unreliable since every soldered joint was a potential source of trouble. The challenge for the industry was to find a cost-effective and reliable way of producing these components and interconnecting them.

Jack Kilby (Fig. 3.16) joined Texas Instruments in 1958, and he began investigating how to solve this problem. He realised that semiconductors were all that were really required, as resistors and capacitors could be made from the same

Fig. 3.16 Jack Kilby c.
1958. Courtesy of Texas
Instruments

material as the transistors. He realised that since all the components could be made of a single material that they could also be made in situ interconnected to form a complete circuit.

Kilby succeeded in building an integrated circuit made of germanium that contained several transistors in 1958, and Robert Noyce of Fairchild Semiconductors built an integrated circuit on a single wafer of silicon in 1960. Kirby and Noyce are considered co-inventors of the integrated circuit, and Kilby was awarded the Nobel Prize in Physics in 2000 for his role in its invention.

Kilby's integrated circuit consisted of a transistor and other components on a slice of germanium (Fig. 3.17). His invention revolutionised the electronics industry, and the integrated circuit is the foundation of almost every electronic device in use today. His integrated circuit was 7/16 by 1/16-in.

Robert Noyce at Fairchild Semiconductors developed an integrated circuit based on a single wafer of silicon in 1960, and today silicon is the material of choice for semiconductors. Noyce made an important improvement on Kirby's design, in that he added a thin layer of metal to the chip to better connect the various components in the circuit. Noyce's solution made the integrated circuit more suitable for mass production, and Fairchild Semiconductors pioneered the use of the *planar process* for making transistors. The existing semiconductor companies soon employed this process. Noyce was one of the co-founders of Intel, which is one of the largest manufacturers of integrated circuits in the world.

Fig. 3.17 First integrated circuit. Courtesy of Texas Instruments

An integrated circuit (IC) consists of a set of electronic circuits on a small chip of semiconductor material, and it is much smaller than a circuit made from independent components. The IC is made on a small plate of semiconductor material that is usually made of silicon. It is extremely compact, and it may contain billions of transistors and other electronic components in a tiny area. The width of each conducting line has got smaller and smaller due to advances in technology over the years, and it is now measured in tens of nanometres.[5] The invention of the integrated circuit led to major reductions in the size and cost of making electronics, and it impacted the design of all future computers and other electronics.

The size of the components in a modern fabrication plant is extremely small, with thousands of transistors fitting inside the cross section of a strand of hair. The production of a chip requires precision at the atomic level, with tiny particles such as those in tobacco smoke are large enough to ruin a chip. For this reason, chip production takes place in a clean room, which is a special room designed with furniture made of special materials that don't give off particles, and very effective air filters and air circulation systems.

There has been a massive reduction in the production costs of integrated circuits, with the initial production cost of integrated circuits at \$1000 in 1960. However, as demand increased and production techniques improved, the cost of production was reduced to \$25 by 1963.

There are several generations of integrated circuits, from the small-scale integration (SSI) of the early 1960s, which typically had less than 30 transistors on the chip, to medium scale integration (MSI) of the late 1960s with less than 300 transistors on the chip; to large-scale integration (LSI) of the mid-70s with less than 3000 transistors on the chip; to very large-scale (VLSI) and ultra large-scale integration (ULSI) of the 1980s; which have over a million transistors on the chip. For more detailed information on Jack Kirby and Texas Instruments see O'Regan (2013).

3.4.2 Moore's Law

Gordon Moore observed that over a period of time (from 1958 up to 1965) that the number of transistors on an integrated circuit doubled approximately every year. This led him to formulate what became known as *Moore's Law* in 1965 (Moore 1965), which predicted that this trend would continue for at least another ten years. He refined the law in 1975 and predicted that a doubling in transistor density would occur every two years for the following ten years.

His prediction of *exponential growth* in transistor density has proved to be accurate over the last 50 years, and the capabilities of many digital electronic devices are linked to Moore's Law.

---

[5]1 nm (nm) is equal to 10^{-9} m.

The exponential growth in the capability of processor speed, memory capacity and so on is all related to this law. It is likely that the growth in transistor density will slow down to a doubling of density every 3 years.

The phenomenal growth in productivity is due to continuous innovation and improvement in manufacturing processes. It has led to more and more powerful computers running more and more sophisticated applications.

3.4.3 Early Integrated Circuit Computers

It took some time for integrated circuits to take off, as they were an unproven technology and they remained expensive until mass production. Kilby and others at Texas Instruments successfully commercialised the integrated circuit by designing a handheld calculator that was as powerful as the existing large, electromechanical desktop models. The resulting electronic handheld calculator was small enough to fit in a coat pocket. This battery-powered device could perform the four basic arithmetic operations on six-digit numbers, and it was completed in 1967.

The earliest computers that used integrated circuits appeared in the 1960s, and the early use of integrated circuits was mainly in embedded systems. They played an important role in early aerospace projects such as the Apollo Guidance Computer and Minuteman missile. The Apollo flight computer was one of the earliest computers to use integrated circuits, and it was developed by MIT/Raytheon and introduced in 1966. It provided capabilities for the guidance, navigation and control of the Apollo spacecraft. The Minuteman II program used a computer built from integrated circuits, and the guidance system of the Minuteman II intercontinental ballistic missile was much smaller due to their use.

DEC's first minicomputer to use integrated circuits was the popular PDP-8 (Fig. 3.18), which was designed by Edson de Castro, and introduced in 1965. Hewlett-Packard introduced the 2116A minicomputer in 1966, and this minicomputer used Fairchild Semiconductors integrated circuits.

The Honeywell ALERT airborne computer was designed to handle complex airborne data in a real-time environment, and it was introduced in 1966. The Central Air Data Computer, was designed in the late 1960s, and it was used for flight control in the US Navy's F-14A Tomcat Fighter.

3.4.4 Birth of Silicon Valley

Silicon Valley is the nickname for the southern portion of the San Francisco Bay area, and it is home to many of the world's largest high-tech companies as well as thousands of start-up companies.

Fig. 3.18 The DEC PDP-8/e

The term '*Silicon Valley*' first appeared in the printed media in 1971, in a series by Don Hoefler titled '*Silicon Valley in the USA*', which was published in the weekly newspaper *Electronics News*. The term was used widely from the early 1980s following the introduction of the IBM personal computer, and the high concentration of semiconductor technology companies in the area. The word '*silicon*' originally referred to the large number of silicon chip manufacturers in the area, as most semiconductors are made from silicon. The word '*valley*' refers to the Santa Clara Valley.

Bill Hewlett and Dave Packard started their two-person company (Hewlett-Packard) in a Palo Alto garage (Fig. 3.19) on 367 Addison Street in 1938. Fruit orchards covered the surrounding area, as Silicon Valley, as it is known today, did not exist. This 12 by 18-foot garage is now a historical landmark, and it has been officially declared the '*birthplace of Silicon Valley*'. HP purchased the property in 2000 to preserve it for future generations.

William Shockley (one of the inventors of the transistor) moved from New Jersey to Mountain View in California to start Shockley Semiconductors in 1956. Shockley's work served as the foundation for many electronics developments, but his management style led to the resignation of eight key researchers in 1957 (the '*traitorous eight*'). This gang of eight went on to form Fairchild Semiconductors and other companies in the Silicon Valley area in the following years. They included Gordon Moore and Robert Noyce, who founded Intel in 1968. Other employees from Fairchild Semiconductors formed companies such as National Semiconductors and Advanced Micro Devices in the Silicon Valley area in later years. Shockley Semiconductors and these new companies formed the nucleus of what became Silicon Valley.

Fig. 3.19 HP Palo Alto Garage. Birthplace of Silicon Valley. Courtesy of HP

Stanford University played an important role in the development of Silicon Valley, and Frederick Terman, the Dean of Engineering and provost of Stanford University in the 1950s, encouraged graduates to form companies in the Silicon Valley area. Stanford University set up an industrial park (Stanford Research Park) for high-technology companies. Terman has been described as the father of Silicon Valley.

3.5   IBM System 360

The IBM System/360[6] was a family of mainframe computers designed and developed by IBM. It set IBM on the road to dominate the computing field for the next twenty years, up to the introduction of personal computers in the 1980s. It was the beginning of an era of computer compatibility, where machines across a product line could work with each other. It meant that IBM customers could start off with a low specification member of the family, and upgrade over time to a more powerful member.

This allowed the customer to choose the appropriate model to meet its current needs, and it could upgrade to a more powerful member of the family as its needs evolved. It was a massive $5 billion investment (*bet the business gamble*) by Thomas Watson Jr., and it moved IBM from its traditional business and product lines into the unknown with the gamble that the future would be the System/360.

---

[6]The number '360' (the number of degrees in a circle) was chosen to represent the ability of each computer to handle all types of applications.

Thomas Watson Jr.[7] announced the System/360 in 1964, and it changed business and the world of computing. The System 360 replaced all five of IBM's computer product lines with one strictly compatible family. It used a new computer architecture that employed hybrid integrated circuit technology, and it pioneered the 8-bit byte, which remains in use on every computer today.

The System/360 included a multiprogramming disk-based operating system, which was called OS/360. It included free software packages such as compilers for several programming languages, as well as packages for communication network capabilities (Pugh 2009).

It was an extremely successful product line for IBM with orders rapidly exceeding forecasts and over a thousand orders placed in the first four weeks after the announcement. Its success made it difficult for IBM competitors[8] (such as Burroughs, Honeywell and Sperry-Rand) to compete against IBM in the general-purpose computer market.

Monthly rental prices ranged from under $3000 per month for the most basic system to over $100,000 per month for a large system. The purchase cost ranged from $130,000 for a basic system to over $5 million for a large system. In 1989, 25 years after the announcement of the System/360, products based on the System/360 architecture and its extensions still accounted for over 50% of IBM revenue.

3.5.1   Background to the Development of System/360

Thomas Watson Jr., the son of Thomas Watson Sr. (the first president of IBM), became president of IBM in 1952. He recognised that computers would play a key role for business in the years ahead, and he realised that the future of IBM was in the computer business, and not in tabulators. It was clear to him that IBM needed to change, and he played a key role in transforming the company to become the world leader in the computer industry.

IBM was already a successful computer company in the 1950s. It introduced its first large computer (the IBM 701) based on vacuum tubes in 1952; the IBM 650 (Magnetic Drum Calculator) in 1954, and the IBM 704 data processing system computer in 1954. It had also played a key role in the development of the computers for the SAGE air defence system in the United States. IBM was the market leader and it employed over 100,000 people around the world. That is, IBM was the 'Snow White' of the computer industry, and Burroughs, Sperry, NCR, Control Data Corporation, Honeywell, General Electric and RCA were the seven dwarfs of the computer sector.

---

[7]Thomas Watson Jr. later stated, 'The System/360 was the biggest, riskiest decision that I ever made, and I agonised about it for weeks, but deep down I believed that there was nothing that IBM couldn't do'.

[8]IBM and its competitors were referred to as Snow White and the Seven Dwarfs.

However, within IBM there were concerns that the company had reached a plateau, and competitors were launching alternative products to IBM. The origins of the System/360 go back to the late 1950s, and Watson's determination to transform IBM to position it for future success. IBM was supporting five different product lines by 1959, and it was becoming a major challenge to train staff to service and maintain software to support so many different computer products.

There were major problems with incompatibility between different hardware and software among the different computer vendors, as well as incompatibility among IBM's own products. IBM had an existing product line of several computers, each excellent, but all with incompatible architectures. It meant that customers who wished to move up from their existing small system to a larger system had to invest in a new system, new printers, new storage devices and new software (often totally rewritten for the new machine).

It was clear to Watson and other senior IBM executives that there was a need to develop a totally cohesive product line so that computers produced at different IBM facilities would be compatible with one another. IBM set up a corporate-wide task group to establish an overall IBM plan for its future products. The task group had the acronym SPREAD (System Programming, Research, Engineering and Design), and it completed its final report in late 1961. It made a series of recommendations such as that there would be five processors spanning a 200-fold range in performance. IBM made the brave decision in 1962 to replace the company's entire product line of computers, and to build a new family of compatible machines (Fig. 3.20).

Fig. 3.20 IBM System/360. Courtesy of International Business Machines Corporation, © International Business Machines Corporation

It would mean that code written for the smallest member of the family would be upwardly compatible with each of the processors in the family. Further, the various peripherals such as printers and storage devices would be compatible across the family. It was an incredibly brave decision, and Fortune Magazine later described it as *'IBM's five-billion-dollar gamble'*.

3.6 Minicomputers and Later Mainframes

The *minicomputer* was a new class of low-cost computers that arose during the 1960s. The development of minicomputers was facilitated by the introduction of integrated circuits, and their improved performance and declining cost. Minicomputers were distinguished from the large mainframe computers by price and size, and they formed a class of the smallest general-purpose computers.

Mainframes were large, expensive machines (typically costing over $1 million) and they required separate rooms for technicians and operation, whereas minicomputers cost well under $100,000 and they were designed for direct, personal interaction with the programmer.

Digital Equipment Corporation (DEC) and Control Data Corporation (CDC) introduced small or minicomputers in the early 1960s. These included DEC's PDP-1, which was released in 1961, and the CDC-160A, which was released in 1960. These machines cost $110,000 and $60,000, respectively, which was a fraction of the cost of a mainframe computer.

The DEC PDP series of minicomputers became popular in the 1960s. The PDP-8 minicomputer (Fig. 3.18) was released in 1965, and it was a 12-bit machine with a small instruction set. The PDP-11 (Fig. 3.21) was a family of 16-bit minicomputers produced by DEC from 1970 to the early 1990s. It was DEC's most successful computer, with over 600,000 machines sold. It was the only 16-bit computer made by the company, as its successor was the 32-bit VAX:11 series. It started its life as a minicomputer and ended its life as micro/super-microcomputer. The rise of the microprocessor and microcomputer led to the availability of low-cost personal computers, and this later challenged DEC's product line. DEC was too late in responding to the paradigm shift in the industry, and this proved to be fatal for the company. Compaq acquired DEC in 1998 for $9.8 billion, and HP later acquired Compaq.

Gene Amdahl was the chief architect for the IBM System/360, and he resigned from IBM to set up Amdahl Corporation in 1970. His goals were to develop a mainframe that would provide better performance than the existing IBM machines, and do so a lower cost, as well as being compatible with IBM hardware and software. Amdahl Corporation launched its first product, the Amdahl 470V/6, in 1975. This was an IBM S/370 compatible mainframe that could run IBM software, and so it was an alternative to a full IBM proprietary solution. It meant that companies around the world now had the choice of continuing to run their software on IBM machines or purchasing the cheaper and more powerful IBM compatibles

Fig. 3.21 PDP-11
minicomputer

produced by Amdahl. Amdahl Corporation became a major competitor to IBM in
large-scale computer placements.

IBM launched a new product, the IBM 3033, in 1977 to compete with the
Amdahl 470. However, Amdahl Corporation responded with a new machine, the
470V/7, which was one and a half times faster than the 3033, and only slightly
more expensive. Customers voted with their feet and chose Amdahl as their supplier, and by late 1978 it had sold over a hundred of the 470V/7 machines.

IBM introduced a medium-sized computer, the 4300 series, in early 1979, and in
late 1980 it announced plans for the 3081 processor, which would have twice the
performance of the existing 3033 on its completion in late 1981. In response, Amdahl
announced the 580 series that would have twice the performance of the existing 470
series. The 580 series was released in mid-1982, but their early processors had some
reliability problems and lacked some of the features of the new IBM product.

Amdahl moved into large system multiprocessor design from the mid-1980s. It
introduced its 5890 model (Fig. 3.22) in late 1985, and its superior performance
allowed Amdahl to gain market share and increase its sales to approximately
$1 billion in 1986. It now had over 1300 customers in around twenty countries
around the world. It launched a new product line, the 5990 processor, in 1988, and
this processor outperformed IBM by 50%. Customers voted with their feet and
chose Amdahl as their supplier.

It was clear that Amdahl was now a major threat to IBM in the high-end
mainframe market. Amdahl had a 24% market share and annual revenues of
$2 billion at the end of 1988. This led to a price war with IBM, with the latter
offering discounts to its customers to protect its market share. Amdahl responded
with its own discounts, and this led to a reduction in profitability for the company.

The IBM personal computer was introduced in the early 1980s, and by the early
1990s, it was clear that the major threat to Amdahl was the declining mainframe

Fig. 3.22 Amdahl 5890. Courtesy of Robert Broughton, University of Newcastle

market rather than IBM. Revenue and profitability fell, and Amdahl shut factory lines and cut staff numbers. By the late 1990s, Amdahl was making major losses, and there were concerns about the future viability of the company. Amdahl became a wholly owned subsidiary of Fujitsu in 1997, and it exited the mainframe business in 2002.

3.7 Microprocessor Revolution

The invention of the microprocessor (initially called microcomputer) in 1971 was a revolution in computing. It meant that the power of a computer was now available on a tiny microprocessor chip, and its invention made handheld calculators and personal computers (PCs) possible. Intel's microprocessors are used on most PCs and laptops around the world.

A *microprocessor* is a central part of a modern personal computer (or computer device). It integrates the functions of the central processing unit (the part of a computer that processes the program instructions) onto a single integrated circuit, and places a vast amount of processing power in a tiny space.

The microprocessor was initially developed as an enhancement to allow users to add more memory to their units. However, it soon became clear that it had great potential for everything from calculators to cash registers and traffic lights. Its invention made personal computers, tablets and mobile phones possible.

The invention of the microprocessor happened by accident rather than design. The Nippon Calculating Machine Corporation (later known as Busicom), a Japanese company, requested Intel to design a set of integrated circuits for its new family of high-performance programmable calculators. At that time, it was standard practice to custom design all logic chips for each customer's product, and this clearly limited the applicability of a logic chip to a specialised domain.

The design proposed by Busicom required twelve integrated circuits. Ted Hoff, an Intel engineer, studied Busicom's design and he rejected it as unwieldy. He proposed a more elegant solution requiring just four integrated circuits, and his design included a chip that was a general-purpose logic device (microprocessor) that derived its application instructions from the semiconductor memory. Busicom accepted his proposed design, and Intel engineers then implemented it.

Hoff's 4004-microprocessor design (Fig. 3.23) included a central processing unit (CPU) on one chip. It contained 2300 transistors on a one-eighth by one-sixth-inch chip surrounded by three ICs containing ROM, shift registers, input/output ports and RAM.

Busicom had exclusive rights to the design and components, but following discussion and negotiations, Busicom agreed to give up its exclusive rights to the chips. Intel shortly afterwards announced the availability of the first microprocessor, the Intel 4004.

This was the world's first microprocessor, and although it was initially developed as an enhancement to allow users to add more memory to their units, it soon became clear that the microprocessor could be applied to many other areas.

This small Intel 4004 microprocessor chip was launched in late 1971, and it could execute 60,000 operations per second. The tiny chip had an equivalent computing power as the large ENIAC computer, which used 18,000 vacuum tubes, and took up the space of an entire room (O'Regan 2016a).

The Intel 4004 sold for $200 and for the first-time affordable computing power was available to designers of all types of products. The introduction of the microprocessor was a revolution in computing, and its invention had applications to everything from traffic lights, to medical instruments, and to the development of home and personal computers.

Fig. 3.23 Intel 4004
microprocessor

Gary Kildall was one of the early people to recognise the potential of the microprocessor as a computer in its own right, and he began writing experimental programs for the Intel 4004 in the early 1970s. Kildall worked as a consultant with Intel on the later 8008 and 8080 microprocessors.

He developed the first high-level programming language for a microprocessor (PL/M) in 1973, which enabled programmers to write applications for microprocessors. He developed the CP/M operating system (Control Program for Microcomputers), which allowed the Intel 8080 microprocessor to control a floppy disk drive allowing files to be read and written to and from an eight-inch floppy disk. CP/M made it possible for computer hobbyists and companies to build the first home computers.

Kildall made CP/M hardware independent by creating a separate module called the BIOS (Basic Input/Output System). He added several utilities such as an editor, debugger and assembler, and by 1977 several manufacturers were including CP/M with their systems. He set up Digital Research Inc. (DRI) in 1976 to commercialise the CP/M operating system.

3.7.1   Early Microprocessors

Intel has introduced more and more powerful microprocessors since its invention of the Intel 4004 in 1971. The Intel 8008 was launched in 1972, and it led to the 8-bit Intel 8080 microprocessor, which was released in 1974. The Intel 8080 was the first general-purpose microprocessor, and it was sold for $360, i.e. a whole computer on one chip was sold for $360, while conventional computers were sold for hundreds of thousands of dollars. The Intel 8080 soon became the industry standard, and Intel became the industry leader in the 8-bit market. The 8080 played an important role in starting personal computer development, as it attracted the interest of computer developers and engineers.

IBM considered several microprocessors for its IBM PC including the IBM 801 processor, the Motorola 68000 microprocessor, and the Intel 8088 microprocessor. IBM chose the Intel 8088 chip (which was cheaper than the 16-bit Intel 8086), and it took a 20% stake in Intel leading to strong ties between both companies.

Today, Intel's microprocessors are used on most personal computers around the world, and the contract to supply the Intel 8088 microprocessor was a major turning point for the company. Intel had been focused more on the sale of dynamic random-access memory chips, with sales of microprocessors in thousands or in tens of thousands. However, sales of microprocessors rocketed following the introduction of the IBM PC, and soon sales were in tens of millions of units.

The cost of computing processing power has fallen exponentially since the introduction of the first microprocessor, and Intel has played a key role in squeezing more and more transistors onto a chip leading to more and more powerful microprocessors and personal computers.

3.8 Home Computers

The invention of the microprocessor was a revolution in computing, and it led to the development of home and personal computers. We discuss a selection of home computers in this section, including the MITS Altair 8800; the Apple I and II computers; the Commodore PET and Commodore 64 computers; and the Atari 400 and 800 computers. We discuss the Apple Macintosh computer, which was a major milestone in computing.

Many of the early home computers were based on the inexpensive 8-bit MOS 6502 microprocessor (e.g. Apple I, the Atari 400 and the Commodore PET). The MITS Altair 8800 was based on the Intel 8080 microprocessor, and later home and personal computers used a variety of microprocessors. The Apple Macintosh was based on the Motorola 68000 microprocessor, and the Atari personal computer was based on the Intel 8088 microprocessor.

We start with a discussion of the Xerox Alto computer, which was developed at Xerox PARC. It pioneered several key concepts in personal computing and had a major impact on the design of the Apple Macintosh.

3.8.1 Xerox Alto Personal Computer

The Xerox Alto (Fig. 3.24) was one of the earliest personal computers, and it was introduced in early 1973. Chuck Thacker and others at Xerox designed it, and it was one of the first computers to use a mouse-driven graphical user interface. It was designed for individual rather than home use and was used by a single person sitting at a desk. It was essentially a small minicomputer rather than a personal computer, and it was unlike modern personal computers in that it was not based on the microprocessor. The significance of the Xerox Alto is that it had a major impact on the design of early personal computers, and especially on the design of the Apple Macintosh computer.

3.8.2 MITS Altair 8800

Micro Instrumentation and Telemetry Systems (MITS) was founded by Ed Roberts and others in 1969. Roberts had a background in electronics from the US military, and the initial focus of the company was to design and sell electronic kits to model rocket enthusiasts. This had become a popular hobby in the 1960s, due to manned space flights and the race to the moon.

MITS began working on the Altair 8800 home computer (Fig. 3.25) in 1974, and the prototype was available in October of that year. The cover page of the January 1975 edition of Popular Electronics featured an early design of the Altair

Fig. 3.24 Xerox Alto

Fig. 3.25 MITS Altair
Computer. Photo Public
Domain

8800, and this publicity helped in generating sales that vastly exceeded expectations. Over 5000 machines were delivered by August 1975, and the home computer kit version (which was assembled by the customer) cost $439, whereas the fully assembled version cost $621.

The home kit included assembly instructions, a metal case, a front panel with switches, a power supply, a motherboard with expansion slots and various cards to plug into the expansion slots, as well as any other components required to build the computer. The actual assembly was quite a challenge as it involved careful soldering and assembly. There was no actual keyboard or monitor, which meant that the task of programming the machine was non-trivial, and required the user to program in machine language and watch the LEDs on the front panel to get the results. Several expansion cards (e.g. for keyboard, monitor and data storage) were soon released, and this made it easier to use. The Altair 8800 used the 8-bit Intel 8080 microprocessor, which was introduced in 1974.

Bill Gates and Paul Allen developed a BASIC interpreter for the Altair 8800, and the 4k/8k versions of BASIC were released in July 1975. This cost the customer an additional $60/$75. Gates and Allen formed Microsoft later in 1975, and Altair BASIC was their first product.

3.8.3 Apple I and II Home Computers

Steve Jobs and Steve Wozniak formed Apple Computer, Inc. in 1976, and the company commenced operations in Jobs' family garage. Their goal was to develop a user-friendly alternative to the existing mainframe and minicomputers produced by IBM and Digital. Wozniak was responsible for product development and Jobs for marketing. Jobs and Wozniak were both college dropouts, and both attended the Homebrew Computer Club of computer enthusiasts in Silicon Valley during the mid-1970s.

The Apple I computer was released in 1976, and it was mainly of interest to computer hobbyists and engineers. This was since it was not a fully assembled personal computer as such, and it was essentially an assembled motherboard that lacked features such as a keyboard, monitor and case. It used a television as the display system, and it had a cassette interface to allow programs to be loaded and saved. It used the inexpensive MOS Technologies 6502 microprocessor chip, which had been released earlier that year, and Wozniak had already written a BASIC interpreter for this chip.

The Apple II computer (Fig. 3.26) was released in 1977, and it was a significant advance on its predecessor. It was a personal computer with a monitor, keyboard and case, and it was one of the earliest computers to come pre-assembled. It was a popular 8-bit home computer, and it was one of the earliest computers to have a colour display with colour graphics.

The BASIC programming language was built in, and it contained 4K of RAM (which was could be expanded to 48K). The VisiCalc spreadsheet program was released on the Apple II, and this helped to transform the computer into a credible business machine. For more detailed information on Apple see O'Regan (2015).

3.8.4 Commodore PET

Commodore Business Machines was a leading North American home computer and electronics manufacturing company. It played an important role in the development of the home computer industry in the 1970s and 1980s, and it is especially famous for its development of the Commodore PET computer, which was very popular in the education field. It also developed the popular VIC-20 and Commodore 64 home computers.

Fig. 3.26 Apple II
Computer. Photo Public
Domain

Commodore initially manufactured typewriters for the North American market, and it diversified into the manufacture of mechanical calculators from the early 1960s. It introduced both consumer and scientific calculators in the late 1960s, and by the early 1970s it was one of the most popular brands. The calculators used Texas Instruments chips but when Texas Instruments entered the calculator market in the mid-1970s, Commodore was unable to compete with their prices.

Commodore purchased the semiconductor company, MOS technologies, with the intention of using MOS chips in its calculators. However, Chuck Peddle, one of MOS's employees convinced Commodore that the future was in computers and not calculators. Commodore used one of MOS's Technologies chips, the 8-bit 6502, to enter the home computer market in 1977 with the launch of its Commodore Personal Electronic Transactor (PET) computer.

This Commodore PET was very popular in the education market and one of its models was called the '*Teachers PET*'. It used the MOS 8-bit 6502 microprocessor, which was designed by Chuck Peddle and others at MOS Technology. The 6502 controlled the screen, keyboard, the cassette recorder and any peripherals connected to the expansion ports. The machine used the Commodore BASIC operating system. There were several models of the Commodore PET introduced during its lifetime including the PET 2001 series, the PET 4000 series and the Super PET 8000 series.

The first model introduced was the PET 2001 (Fig. 3.27), which had either 4 Kb or 8 Kb of RAM. It had a built-in monochrome monitor with 40×25 character graphics enclosed in a metal case. It included a magnetic data storage device known as a datasette (data + cassette) in the front of the machine as well as a small keyboard. There were complaints with respect to the small keyboard, which soon led to the appearance of external replacement keyboards.

Fig. 3.27 Commodore PET
2001 Home Computer

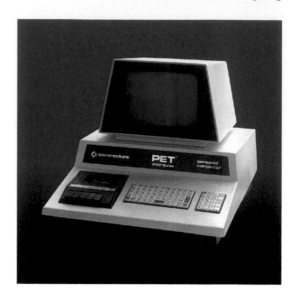

Commodore introduced other models such as the PET 4000-series and the 8000-series. For more detailed information on Commodore see O'Regan (2015).

3.8.5 Atari 400 and 800

Atari designed and produced four lines of home and personal computers from the late 1970s up to the early 1990s. These were the 8-bit Atari 400 and 800 lines, the 16-bit ST line, the IBM PC compatible series and the 32-bit series.

The Atari 8-bit series began as a next-generation follow up to its successful Atari 2600 video game console. Atari's management noted the success of Apple in the early personal computer market, and they tasked their engineers to transform the hardware into a personal computer system. The net result was the Atari 400 and the Atari 800 home computers, which were introduced in 1979.

The Atari 800 (Fig. 3.28) came with 8 KB of RAM and the Atari 400 was a lower specification version. Both machines were based on the MOS 6502 microprocessor, and they provided sound and graphics capabilities that were superior to competitor products such as the Apple II or the Commodore PET.

The Atari 400 and 800 made an impact on the home computing field, and both machines included joystick ports for playing games. Atari BASIC was provided on an external cartridge for each machine.

Fig. 3.28 The Atari 800
Home Computer

The Atari 400 was Atari's entry-level computer, and it was designed for younger children. It had a membrane keyboard designed to prevent damage from food or small objects, and the keys could not be removed or swallowed by children. It was initially designed for 4K of memory but as memory costs declined it was shipped with 8K (and later 16K). This meant that it could run most cartridge- and cassette-based software. It was connected to a standard television.

The Atari 800 came with a graphics/audio chipset that allowed it to produce the most advanced graphics and sound on an existing home computer system. For more detailed information on Atari see O'Regan (2015).

3.8.6 Commodore 64

The Commodore 64 (C64) was a very successful 8-bit home computer introduced by Commodore in 1982 (Fig. 3.29). It used the MOS 6501 microprocessor and it came with 64 kbyte of RAM. It had 320×200 colour graphics with 16 colours using the VIC-II graphics chip, and the MOS Sound Interface Device (SID) chip. The SID chip was one of the first sound chips to be included in a home computer. The C64 dominated the low-end home computer market for most of the 1980s, and approximately 15 million of the Commodore 64 machines were sold.

It came with the Commodore BASIC, but support for other languages was also available. The Commodore 64's graphics and sound capabilities were quite advanced for the time, and the C64 was very popular for computer games. Commodore published detailed technical documentation to assist programmers and

Fig. 3.29 Commodore 64
Home Computer

enthusiastic users to develop applications, and this led to over 10,000 commercial software applications such as development tools, games and office productivity applications for the machine.

The C64 included a ROM-based version of the BASIC 2.0 programming language. There was no operating system as such, and instead, the kernel was accessed via BASIC commands. BASIC did not provide commands for sound or graphics manipulation, and instead, the user had to use the 'POKE' command to access these chips directly. The Commodore 64 remained highly popular throughout the 1980s, and there is a more detailed account of Commodore in Bagnall (2012).

3.8.7 Apple Macintosh

The Apple Macintosh (Fig. 3.30) was announced during a famous television commercial played during the third quarter of the Super Bowl in January 1984. This was one of the most creative advertisements of all time, and it ran just once on television. It generated more excitement than any other advertisement up to then, and it immediately positioned Apple as a creative and innovative company, while implying that its competition (i.e. IBM) was stale and robotic.

It presented Orwell's totalitarian world of 1984, with a lady runner with orange shorts and a white tee shirt with a picture of the Apple Macintosh running towards a big screen, and hurling a hammer at the big brother character on the screen. The audience is stunned at the broken screen and the voice over states '*On January 24th Apple will introduce the Apple Macintosh and you will see why 1984 will not be like "1984"*'. Ridley Scott who has directed well-known films such as Alien, Blade Runner, Robin Hood and Gladiator directed the short film.

Fig. 3.30 Apple Macintosh Computer. Photo Public Domain

The Macintosh project began in Apple in 1979 with the goal of creating an easy-to-use low-cost computer for the average consumer. It was influenced by the design of the Apple Lisa, and it employed the Motorola 68000 processor. Steve Jobs became involved in the project in 1981, and he negotiated a deal with Xerox that allowed him and other Apple employees to visit the Xerox PARC research centre at Palo Alto in California to see their pioneering work on the Xerox Alto computer. PARC's research work had a major influence on the design and development of the Macintosh, as Jobs was convinced that future computers would use a graphical user interface. The design of the Macintosh included a friendly and intuitive graphical user interface (GUI), and the release of the Macintosh was a major milestone in computing.

The Macintosh was a much easier machine to use than the existing IBM PC. Its friendly and intuitive graphical user interface was a revolutionary change from the command-driven operating system of the IBM PC, which required the users to be familiar with its operating system commands. The introduction of the Mac GUI is an important milestone in the computing field, and it was 1990 before Microsoft introduced its Windows 3.0 GUI-driven operating system.

Apple intended that the Macintosh would be an inexpensive and user-friendly personal computer that would rival the IBM PC and compatibles. However, it was significantly more expensive than the IBM PC, and initially it had a limited number of applications available, whereas the IBM PC had spreadsheets, word processors and databases applications, and so it was more attractive to customers. The technically superior Apple Macintosh was unable to break the IBM dominance of the market. However, the machine became very popular in the desktop publishing market, due to its advanced graphics capabilities.

3.9 The IBM Personal Computer

The introduction of the IBM personal computer was a paradigm shift in computing with computing power placed in the hands of millions of people. The previous paradigm was that an individual user had limited control over a computer, with the system administrators controlling the access privileges of the users. Today's personal computers are more powerful than the mainframes that were used to send man to the moon.

IBM introduced the IBM Personal Computer (PC) in 1981 as a machine to be used by small businesses and users in the home. The IBM goal at the time was to get quickly into the home computer market, which was then dominated by Commodore, Atari and Apple.

IBM assembled a small team of twelve people led by Don Estridge (Fig. 3.31), and their objective was to get the personal computer to the market as quickly as possible. They designed and developed the IBM PC within one year, and as time to market was the key driver they built the machine with '*off-the-shelf*' parts from several equipment manufacturers.

Fig. 3.31 Don Estridge.
Courtesy of International
Business Machines
Corporation, © International
Business Machines
Corporation

IBM's traditional approach up to then in product development was to develop a full proprietary solution. However, due to the aggressive timescales associated with the introduction of the IBM PC, it decided instead to outsource the development of the microprocessor to a small company called Intel and to outsource the development of the operating system to a small company called Microsoft.

The team had intended using the IBM 801 processor, which was developed at the IBM Research Centre in Yorktown Heights. However, they decided instead to use the Intel 8088 microprocessor, which was inferior to the IBM 801. They chose the PC/DOS operating system from Microsoft rather than developing their own operating system. These decisions would later prove costly to IBM, as Microsoft and Intel reaped the benefits and later became technology giants.

The unique IBM elements in the personal computer were limited to the system unit and keyboard. The team decided on an open architecture so that other manufacturers could produce and sell peripheral components and software without purchasing a license. They published the *IBM PC Technical Reference Manual*, which included the complete circuit schematics; the IBM ROM BIOS source code, and other engineering and programming information.

The IBM PC (Fig. 3.32) was the cheapest IBM computer produced up to then, and it was priced at an affordable $1565. It offered 16 kbyte of memory (expandable to 256 kbyte), a floppy disk, a keyboard and a monitor. The IBM personal computer became an immediate success, and it became the industry standard.

The open architecture led to a new industry of '*IBM-compatible*' computers, which had all the essential features of the IBM PC, except that they were cheaper.

The development of the IBM PC meant that computers were now affordable to ordinary users, and this led to a huge consumer market for personal computers and software. It led to the development of business software such as spreadsheets and accountancy packages, banking packages, programmer developer tools such as compilers for various programming languages; specialised editors; and computer games.

The introduction of the personal computer was a paradigm shift in computing, and it led to a fundamental change in the way in which people worked. It placed computing power directly in the hands of millions of people, with individual users

Fig. 3.32 IBM Personal
Computer. Courtesy of
International Business
Machines Corporation, ©
International Business
Machines Corporation

having complete control over the machine. The previous paradigm was that the
system administrators strictly controlled the access privileges of the individual
users, and so users had limited control over the computer. The introduction of the
client-server architecture led to the linking of the personal computers (clients) to
larger computers (servers). These servers contained large amounts of data that could
be shared with the individual client computers.

The IBM strategy in developing the IBM personal computer was deeply flawed,
and it cost the company dearly. IBM had traditionally produced all the components
for its machines, but with its open architecture model, any manufacturer could now
produce an IBM compatible machine. IBM had outsourced the development of the
microprocessor chip to Intel, and Intel later became the dominant player in the
microprocessor industry.

The development of the operating system, PC/DOS (PC Disk Operating System)
was outsourced to a small company called Microsoft.[9] This proved to be a major
mistake by IBM, as the terms of the licensing deal with Microsoft were favourable
to the latter, and it allowed Microsoft to sell its own version of the operating system
(i.e. MS/DOS) to other manufacturers as the operating system for the IBM compatibles. The IBM compatibles were cheaper than the IBM PC, and over time came
to dominate the personal computer market. This led inexorably to the rise of the
Microsoft Corporation.

---

[9]Microsoft was founded by Bill Gates and Paul Allen in 1975.

3.9.1 Operating System for IBM PC

Digital Research lost out on the opportunity of a lifetime to supply the operating for the IBM personal computer to IBM, and instead, it was Microsoft that reaped the benefits. Bloomberg Business Week published an article in 2004 describing the background to the development of the operating system for the IBM PC, and the failed negotiations between Digital Research and IBM on the licensing of the CP/M operating system. The article was titled '*The Man who could have been Bill Gates*' (Bloomberg Business Week Magazine 2004).

The IBM team initially asked Bill Gates and Microsoft in Seattle to supply them with an operating system. Microsoft had already signed a contract with IBM to supply a BASIC interpreter for the IBM PC, but they lacked the expertise in operating system development. Gates referred IBM to Gary Kildall at DRI, and the IBM team approached Digital Research with a view to licensing its CP/M operating system.

Digital Research was working on a new version of CP/M for the 16-bit Intel 8086 microprocessor, which had been introduced in 1978. IBM decided to use the lower cost Intel 8088 microprocessor (a slower version of the 8086) for its new personal computer.

IBM and Digital Research failed to reach an agreement on the licensing of CP/M for the IBM PC. The precise reasons for failure are unclear, but some immediate problems arose with respect to the signing of an IBM non-disclosure agreement during the visit. It is unclear whether Kildall met with IBM and whether there was an informal handshake agreement between both parties. However, there was no documented legal agreement between IBM and DRI.

There may also have been difficulties in relation to the amount of royalty payment being demanded by Digital Research, as well as practical difficulties in achieving the required IBM delivery schedule (due to Digital Research's existing commitments to Intel).

Gates offered to provide an operating system (later called PC/DOS) and BASIC to IBM on favourable terms. IBM accepted the offer, and the contract allowed Microsoft to market and sell its version (MS/DOS) of the operating systems on IBM compatibles. Microsoft reaped the benefits, as sales of IBM compatibles soared, and MS/DOS became the dominant operating system for personal computers.

Gates was aware of the work done by Tim Patterson on a simple quick and dirty version of CP/M (called QDOS) for the 8086 microprocessor for Seattle Computer Products (SCP). Gates licensed QDOS for $50,000, and he hired Patterson to modify it to run on the IBM PC for the Intel 8088 microprocessor. Gates then licensed the operating system to IBM for a low per-copy royalty fee.

IBM called the new operating system PC/DOS, and Microsoft retained the rights to MS/DOS, which was used on IBM compatible computers produced by other hardware manufacturers. In time, MS/DOS would become the dominant operating system (eclipsing PC/DOS due to the open architecture of the IBM PC and the rapid growth of clones) leading to the growth of Microsoft into a major corporation.

DRI released CP/M-86 shortly after IBM released PC DOS. Kildall examined PC/DOS, and it was clear to him that it had been derived from CP/M. He was furious and met separately with IBM and Microsoft, but nothing was resolved. Digital Research considered suing Microsoft for copying all the CP/M system calls in DOS 1.0, as it was clear that Patterson's QDOS was a copy of CP/M.

Kildall considered his legal options but his legal advice suggested that as intellectual copyright law with respect to software had only been recently introduced in the United States, that it was not clear what constituted infringement of copyright. There was no guarantee of success in any legal action against IBM, and considerable expense would be involved. Kildall threatened IBM with legal action, and IBM agreed to offer both CP/M-86 and PC-DOS. However, as CP/M was priced at $240 and DOS at $60, few personal computer owners were willing to pay the extra cost. CP/M was to fade into obscurity.

Perhaps, if Kildall had played his hand differently Digital Research could well have been, the '*Microsoft*' of the PC industry. Kildall's delay in developing the operating system gave Patterson the opportunity to create his own version. IBM was under serious time pressures with the development of the IBM PC, and Kildall may have been unable to meet the IBM deadline. This may have resulted in IBM dealing with Gates instead of DRI.

Further, the size of the royalty fee demanded by Kildall for CP/M resulted in very low sales for the DRI product, whereas if a more realistic price had been proposed then DRI may have made some reasonable revenue. Nevertheless, Kildall could justly feel hard done by, and he may have viewed Microsoft's actions as the theft of his intellectual ideas and technical inventions.

3.10 Review Questions

1. Explain what is meant by a 'stored program' computer.
2. Explain the significance of the work done at Bletchley Park.
3. Describe the work of the University of Manchester in early computing.
4. Explain the significance of the transistor in the computing field.
5. Explain the significance of the SAGE system to the computing field.
6. Describe the contributions of IBM to the computing field.
7. Explain the significance of Moore's Law.
8. What is an integrated circuit?
9. Why did IBM decide to develop the System/360?
10. Describe the competition between Amdahl Corporation and IBM.
11. What is the significance of the Intel 4004?
12. What are the main contributions made by Intel to the semiconductor field?
13. What is the significance of the Xerox Alto in the history of computing?
14. Discuss the contributions of DRI to the computing field.

3.11   Summary

This chapter considered some of the earliest computers developed in the United States (ENIAC and EDVAC), Britain (Colossus and Manchester Mark I), and Germany (Z1-Z3). The Second World War led to research into the development of digital computers to determine if they could provide faster methods of computation. The early computers were mainly large bulky machines consisting of several thousand vacuum tubes.

We discussed early commercial computers as well as the SAGE air defence system. We discussed the invention of the transistor, and it led to smaller, faster and more reliable computers. The integrated circuit consists of a set of electronic circuits on a small chip of semiconductor material, and it is much smaller than a circuit made from independent components. Its invention was a revolution in computing, and it shrunk the size and cost of making electronics.

The IBM System/360 was a family of small to large computers, and it was a paradigm shift away from the traditional 'one size fits all' philosophy of the computer industry. The minicomputer was a new class of low-cost computers and their development was facilitated by the introduction of integrated circuits, as this helped to reduce cost and size of computers. Minicomputers formed a class of the smallest general-purpose computers.

Intel's invention of the microprocessor in 1971 changed computing forever, and it placed the power of a computer on a tiny chip. A microprocessor is a central part of a modern personal computer (or computer device), and it places a vast amount of processing power on a tiny chip.

It led to the development of home and personal computers, and many of the early home computers were based on the 8-bit MOS 6502 microprocessor, with later home and personal computers using a variety of microprocessors such as the 8-bit Zilog Z80 microprocessor; the Motorola 68000 microprocessor, the Intel 8088 microprocessor and later, Intel microprocessors.

We discussed various home computers such as the Apple I and II computers; the Commodore PET computer; the Atari 400 and 800 computers; the Commodore 64 computer and the Apple Macintosh computer. The IBM Personal Computer was introduced in 1981 and was a major milestone in the computing field.

Chapter 4
Overview Mathematics in Computing

Key Topics

Sets
Set Operations
Russell's Paradox
Computer Representation of Sets
Relations
Composition of Relations
Functions
Partial and Total Functions
Functional Programming
Number Theory
Automata Theory
Graph Theory

4.1 Introduction

This chapter introduces the essential mathematics for computing and discusses fundamental concept such as sets, relations and functions. Sets are collections of well-defined objects; relations indicate relationships between members of two sets A and B; and functions are a special type of relation where there is exactly (or at most)[1] one relationship for each element $a \in A$ with an element in B.

A set is a collection of well-defined objects that contains no duplicates. The term '*well defined*' means that for a given value it is possible to determine whether or not it is a member of the set. There are many examples of sets such as the set of natural numbers $\mathbb{N}$, the set of integer numbers $\mathbb{Z}$ and the set of rational numbers $\mathbb{Q}$. The natural numbers $\mathbb{N}$ is an infinite set consisting of the numbers $\{1, 2, \ldots\}$. Venn diagrams may be used to represent sets pictorially.

---

[1] We distinguish between total and partial functions. A total function $f : A \rightarrow B\ n : \mathbb{N}$ in A whereas a partial function may be undefined for one or more values in A.

© Springer International Publishing AG, part of Springer Nature 2018
G. O'Regan, *World of Computing*,
https://doi.org/10.1007/978-3-319-75844-2_4

A binary relation $R\ (A,\ B)$ where A and B are sets is a subset of the Cartesian product $(A \times B)$ of A and B. The domain of the relation is A and the co-domain of the relation is B. The notation aRb signifies that there is a relation between a and b and that $(a,\ b) \in R$. An n-ary relation $R\ (A_1,\ A_2,\ ...\ A_n)$ is a subset of $(A_1 \times A_2 \times \cdots \times A_n)$. However, an n-ary relation may also be regarded as a binary relation $R(A,\ B)$ with $A = A_1 \times A_2 \times \cdots \times A_{n-1}$ and $B = A_n$.

Functions may be total or partial. A total function $f : A \rightarrow B$ is a special relation such that for each element $a \in A$ there is exactly one element $b \in B$. This is written as $f(a) = b$. A partial function differs from a total function in that the function may be undefined for one or more values of A. The domain of a function (denoted by **dom** f) is the set of values in A for which the partial function is defined. The domain of the function is A if f is a total function. The co-domain of the function is B.

We introduce topics such as number theory, automata theory and graph theory. Number theory is the branch of mathematics that is concerned with the mathematical properties of the natural numbers and integers. Automata theory is the branch of computer science that is concerned with the study of abstract machines and automata. These include finite-state machines, pushdown automata and Turing machines. Graph theory is a practical branch of mathematics that deals with the arrangements of certain objects known as vertices (or nodes) and the relationships between them. We briefly discuss computability and decidability and more detailed information on mathematics in computing is available in O'Regan (2012, 2017).

4.2 Set Theory

A set is a fundamental building block in mathematics, and it is defined as a collection of well-defined objects. The elements in a set are of the same kind, and they are distinct with no repetition of the same element in the set.[2] Most sets encountered in computer science are finite, as computers can only deal with finite entities. Venn diagrams[3] are often employed to give a pictorial representation of a set, and to illustrate various set operations such as set union, intersection and set difference.

There are many well-known examples of sets including the set of natural numbers denoted by $\mathbb{N}$, the set of integers denoted by $\mathbb{Z}$, the set of rational numbers denoted by $\mathbb{Q}$, the set of real numbers denoted by $\mathbb{R}$ and the set of complex numbers denoted by $\mathbb{C}$.

---

[2]There are mathematical objects known as *multi-sets* or *bags* that allow duplication of elements. For example, a bag of marbles may contain three green marbles, two blue and one red marble.

[3]The British logician, John Venn, invented the Venn diagram. It provides a visual representation of a set and the various set theoretical operations. Their use is limited to the representation of two or three sets as they become cumbersome with a larger number of sets.

Example 4.1 The following are examples of sets:

- The books on the shelves in a library.
- The books that are currently overdue from the library.
- The customers of a bank.
- The bank accounts in a bank.
- The set of natural numbers $\mathbb{N} = \{1, 2, 3, \ldots\}$.
- The Integer Numbers $\mathbb{Z} = \{\ldots, -3, -2, -1, 0, 1, 2, 3, \ldots\}$.
- The non-negative integers $\mathbb{Z}^+ = \{0, 1, 2, 3, \ldots\}$.
- The set of Prime Numbers $= \{2, 3, 5, 7, 11, 13, 17, \ldots\}$.
- The Rational Numbers is the set of quotients of integers.

$$\mathbb{Q} = \{p/q : p, q \in \mathbb{Z} \quad \text{and} \quad q \neq 0\}$$

A finite set may be defined by listing all its elements. For example, the set $A = \{2, 4, 6, 8, 10\}$ is the set of all even natural numbers less than or equal to 10. The order in which the elements are listed is not relevant: i.e. the set $\{2, 4, 6, 8, 10\}$ is the same as the set $\{8, 4, 2, 10, 6\}$.

Sets may be defined by using a predicate to constrain set membership. For example, the set $S = \{n : \mathbb{N} : n \leq 10 \wedge n \bmod 2 = 0\}$ also represents the set $\{2, 4, 6, 8, 10\}$. That is, the use of a predicate allows a new set to be created from an existing set by using the predicate to restrict membership of the set. The set of even natural numbers may be defined by a predicate over the set of natural numbers that restricts membership to the even numbers. It is defined by

$$\text{Evens} = \{x | x \in \mathbb{N} \wedge \text{even}(x)\}.$$

In this example, $\text{even}(x)$ is a predicate that is true if x is even and false otherwise. In general, $A = \{x \in E | P(x)\}$ denotes a set A formed from a set E using the predicate P to restrict membership of A to those elements of E for which the predicate is true.

The elements of a finite set S are denoted by $\{x_1, x_2, \ldots x_n\}$. The expression $x \in S$ denotes that the element x is a member of the set S, whereas the expression $x \notin S$ indicates that x is not a member of the set S.

A set S is a subset of a set T (denoted $S \subseteq T$) if whenever $s \in S$ then $s \in T$, and in this case the set T is said to be a superset of S (denoted $T \supseteq S$). Two sets S and T are said to be equal if they contain identical elements: i.e. $S = T$ if and only if $S \subseteq T$ and $T \subseteq S$. A set S is a proper subset of a set T (denoted $S \subset T$) if $S \subseteq T$ and $S \neq T$. That is, every element of S is an element of T and there is at least one element in T that is not an element of S. In this case, T is a proper superset of S (denoted $T \supset S$).

The empty set (denoted by $\varnothing$ or $\{\}$) represents the set that has no elements. Clearly $\varnothing$ is a subset of every set. The singleton set containing just one element x is denoted by $\{x\}$, and clearly $x \in \{x\}$ and $x \neq \{x\}$. Clearly, $y \in \{x\}$ if and only if $x = y$.

Example 4.2

(i) $\{1,2\} \subseteq \{1,2,3\}$.
(ii) $\varnothing \subset \mathbb{N} \subset \mathbb{Z} \subset \mathbb{Q} \subset \mathbb{R} \subset \mathbb{C}$.

The cardinality (or size) of a finite set S defines the number of elements present in the set. It is denoted by $|S|$. The cardinality of an infinite[4] set S is written as $|S| = \infty$.

Example 4.3

(i) Given $A = \{2, 4, 5, 8, 10\}$ then $|A| = 5$.
(ii) Given $A = \{x \in \mathbb{Z} : x^2 = 9\}$ then $|A| = 2$.
(iii) Given $A = \{x \in \mathbb{Z} : x^2 = -9\}$ then $|A| = 0$.

4.2.1 Set Theoretical Operations

Several set theoretical operations are considered in this section. These include the Cartesian product operation, the power set of a set, the set union operation, the set intersection operation, the set difference operation and the symmetric difference operation.

Cartesian Product
The Cartesian product allows a new set to be created from existing sets. The Cartesian[5] product of two sets S and T (denoted $S \times T$) is the set of ordered pairs

---

[4]The natural numbers, integers and rational numbers are countable sets (i.e. they may be put into a one-to-one correspondence with the natural numbers), whereas the real and complex numbers are uncountable sets.

[5]Cartesian product is named after René Descartes who was a famous seventeenth-century French mathematician and philosopher. He invented the Cartesian coordinates system that links geometry and algebra, and allows geometric shapes to be defined by algebraic equations.

$\{(s, t) \mid s \in S, t \in T\}$. Clearly, $S \times T \neq T \times S$ and so the Cartesian product of two sets is not commutative. Two ordered pairs (s_1, t_1) and (s_2, t_2) are considered equal if and only if $s_1 = s_2$ and $t_1 = t_2$.

The Cartesian product may be extended to that of n sets $S_1, S_2, \ldots, S_n$. The Cartesian product $S_1 \times S_2 \times \cdots \times S_n$ is the set of ordered n-tuples $\{(s_1, s_2,\ldots, s_n) \mid s_1 \in S_1, \ s_2 \in S_2, \ldots, s_n \in S_n\}$. Two ordered n-tuples $(s_1, s_2, \ldots, s_n)$ and $(s_1', s_2', \ldots, s_n')$ are considered equal if and only if $s_1 = s_1', \ s_2 = s_2', \ldots, s_n = s_n'$.

The Cartesian product may also be applied to a single set S to create ordered n-tuples of S: i.e. $S^n = S \times S \times \cdots \times S$ (n-times).

Power Set

The power set of a set A (denoted $\mathbb{P}A$) denotes the set of subsets of A. For example, the power set of the set $A = \{1, 2, 3\}$ has eight elements and is given by

$$\mathbb{P}A = \{\emptyset, \{1\}, \{2\}, \{3\}, \{1,2\}, \{1,3\}, \{2,3\}, \{1,2,3\}\}$$

There are $2^3 = 8$ elements in the power set of $A = \{1, 2, 3\}$ where the cardinality of A is 3. In general, there are $2^{|A|}$ elements in the power set of A.

Theorem 4.1 (Cardinality of Power Set of A)
There are $2|A|$ elements in the power set of A

Proof Let $|A| = n$ then the cardinality of the subsets of A are subsets of size $0, 1, \ldots, n$. There are $\binom{n}{k}$ subsets of A of size k.[6] Therefore, the total number of subsets of A is the total number of subsets of size $0, 1, 2$, up to n. That is,

$$|\mathbb{P}A| = \sum_{k=0}^{n} \binom{n}{k}$$

The binomial theorem states that

$$(1+x)^n = \sum_{k=0}^{n} \binom{n}{k} x^k$$

Therefore, putting $x = 1$ we get that

$$2^n = (1+1)^n = \sum_{k=0}^{n} \binom{n}{k} 1^k = |\mathbb{P}A|$$

Union and Intersection Operations

The union of two sets A and B is denoted by $A \cup B$. It results in a set that contains all of the members of A and of B and is defined by

$$A \cup B = \{r \mid r \in A \text{ or } \in B\}$$

---

[6]Permutations and combinations are discussed in Chap. 5 of O'Regan (2017).

For example, suppose $A = \{1, 2, 3\}$ and $B = \{2, 3, 4\}$ then $A \cup B = \{1, 2, 3, 4\}$. Set union is a commutative operation: i.e. $A \cup B = B \cup A$. Venn diagrams are used to illustrate these operations pictorially.

$$A \cup B \qquad\qquad A \cap B$$

The intersection of two sets A and B is denoted by $A \cap B$. It results in a set containing the elements that A and B have in common and is defined by

$$A \cap B = \{r | r \in A \text{ and } r \in B\}$$

Suppose $A = \{1, 2, 3\}$ and $B = \{2, 3, 4\}$ then $A \cap B = \{2, 3\}$. Set intersection is a commutative operation: i.e. $A \cap B = B \cap A$.

Union and intersection may be extended to more generalised union and intersection operations. For example,

$$\cup_{i=1}^{n} A_i \quad \text{denotes the union of } n \text{ sets}$$
$$\cap_{i=1}^{n} A_i \quad \text{denotes the intersection of } n \text{ sets}$$

Set Difference Operations

The set difference operation $A \backslash B$ yields the elements in A that are not in B. It is defined by

$$A \backslash B = \{a | a \in A \text{ and } a \notin B\}$$

For A and B defined as $A = \{1, 2\}$ and $B = \{2, 3\}$, we have $A \backslash B = \{1\}$ and $B \backslash A = \{3\}$. Clearly, set difference is not commutative: i.e. $A \backslash B \neq B \backslash A$. Clearly, $A \backslash A = \varnothing$ and $A \backslash \varnothing = A$.

The symmetric difference of two sets A and B is denoted by $A \, \Delta \, B$ and is given by:

$$A \, \Delta \, B = A \backslash B \cup B \backslash A$$

The symmetric difference operation is commutative: i.e. $A \, \Delta \, B = B \, \Delta \, A$. Venn diagrams are used to illustrate these operations pictorially.

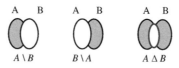

$$A \backslash B \qquad\qquad B \backslash A \qquad\qquad A \, \Delta \, B$$

The complement of a set A (with respect to the universal set U) is the elements in the universal set that are not in A. It is denoted by A^c (or A') and is defined as

$$A^c = \{u | u \in U \text{ and } u \notin A\} = U \backslash A$$

The complement of the set A is illustrated by the shaded area below:

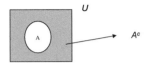

4.2.2 Properties of Set Theoretical Operations

The set union and set intersection properties are commutative and associative. Their properties are listed in Table 4.1.

These properties may be seen to be true with Venn diagrams, and we give a proof of the distributive property (this proof uses logic which is discussed in Chap. 6).

Proof of Properties (Distributive Property)

To show $A \cap (B \cup C) = (A \cap B) \cup (A \cap C)$

Suppose $x \in A \cap (B \cup C)$ then

$$x \in A \wedge x \in (B \cup C)$$
$$\Rightarrow x \in A \wedge (x \in B \vee x \in C)$$
$$\Rightarrow (x \in A \wedge x \in B) \vee (x \in A \wedge x \in C)$$
$$\Rightarrow x \in (A \cap B) \vee x \in (A \cap C)$$
$$\Rightarrow x \in (A \cap B) \cup x \in (A \cap C)$$

Therefore, $A \cap (B \cup C) \subseteq (A \cap B) \cup (A \cap C)$.
Similarly, $(A \cap B) \cup (A \cap C) \subseteq A \cap (B \cup C)$.
Therefore, $A \cap (B \cup C) = (A \cap B) \cup (A \cap C)$.

4.2.3 Russell's Paradox

Bertrand Russell (Fig. 4.1) was a famous British logician, mathematician and philosopher. He was the co-author with Alfred Whitehead of *Principia Mathematica*, which aimed to derive all the truths of mathematics from logic. Russell's paradox was discovered by Bertrand Russell in 1901 and showed that the system of logicism being proposed by Frege (discussed in Chap. 6) contained a contradiction.

QUESTION (POSED BY RUSSELL TO FREGE)

Is the set of all sets that do not contain themselves as members a set?

Table 4.1 Properties of set operations

Property	Description
Commutative	Union and intersection operations are commutative: i.e. $S \cup T = T \cup S$ $S \cap T = T \cap S$
Associative	Union and intersection operations are associative: i.e. $R \cup (S \cup T) = (R \cup S) \cup T$ $R \cap (S \cap T) = (R \cap S) \cap T$
Identity	The identity under set union is the empty set $\emptyset$, and the identity under intersection is the universal set U $S \cup \emptyset = \emptyset \cup S = S$ $S \cap U = U \cap S = S$
Distributive	The union operator distributes over the intersection operator and vice versa $R \cap (S \cup T) = (R \cap S) \cup (R \cap T)$ $R \cup (S \cap T) = (R \cup S) \cap (R \cup T)$
De Morgan's[a] law	The complement of $S \cup T$ is given by $(S \cup T)^c = S^c \cap T^c$ The complement of $S \cap T$ is given by $(S \cap T)^c = S^c \cup T^c$

[a]De Morgan's law is named after Augustus De Morgan, a nineteenth-century English mathematician who was a contemporary of George Boole

Fig. 4.1 Bertrand Russell

RUSSELL'S PARADOX

Let $A = \{S$ a set and $S \notin S\}$. Is $A \in A$? Then $A \in A \Rightarrow A \notin A$ and vice versa. Therefore, a contradiction arises in either case and there is no such set A.

Two ways of avoiding the paradox were developed in 1908, and these were Russell's theory of types and Zermelo set theory. Russell's theory of types was a response to the paradox by arguing that the set of all sets is ill-formed. Russell developed a hierarchy of individual elements: the lowest level, sets of elements at the next level, sets of sets of elements at the next level and so on. It is then prohibited for a set to contain members of different types.

A set of elements has a different type from their elements, and one cannot speak of the set of all sets that do not contain themselves as members as these are of different types. The other way of avoiding the paradox was Zermelo's axiomatization of set theory.

Remark Russell's paradox may also be illustrated by the story of a town that has exactly one barber who is male. *The barber shaves all and only those men in town who do not shave themselves.* The question is who shaves the barber.

If the barber does not shave himself then according to the rule he is shaved by the barber (i.e. himself). If he shaves himself, then according to the rule he is not shaved by the barber (i.e. himself).

The paradox occurs due to self-reference in the statement, and a logical examination shows that the statement is a contradiction.

4.2.4   Computer Representation of Sets

Sets are fundamental building blocks in mathematics, and so the question arises as to how is a set is stored and manipulated in a computer. The representation of a set M on a computer requires a change from the normal view that the order of the elements of the set is irrelevant, and we will need to assume a definite order in the underlying universal set m from which the set M is defined.

That is, a set is always defined in a computer program with respect to an underlying universal set, and the elements in the universal set are listed in a definite order. Any set M arising in the program that is defined with respect to this universal set m is a subset of m. Next, we show how the set M is stored internally on the computer.

The set M is represented in a computer as a string of binary digits $b_1, b_2 \ldots b_n$ where n is the cardinality of the universal set m. The bits b_i (where i ranges over the values 1, 2, ... n) are determined according to the rule:

$b_i = 1$ if ith element of m is in M.
$b_i = 0$ if ith element of m is not in M.

For example, if $m = \{1, 2, \ldots 10\}$ then the representation of $M = \{1, 2, 5, 8\}$ is given by the bit string 1100100100 where this is given by looking at each element of m in turn and writing down 1 if it is in M and 0 otherwise.

Similarly, the bit string 0100101100 represents the set $M = \{2, 5, 7, 8\}$, and this is determined by writing down the corresponding element in m that corresponds to a 1 in the bit string.

Clearly, there is a one-to-one correspondence between the subsets of m and all possible n-bit strings. Further, the set theoretical operations of set union, intersection and complement can be carried out directly with the bit strings (provided that the sets involved are defined with respect to the same universal set). This involves a bitwise 'or' operation for set union, a bitwise 'and' operation for set intersection, and a bitwise 'not' operation for the set complement operation.

4.3   Relations

A binary relation $R(A, B)$, where A and B are sets, is a subset of $A \times B$: i.e. $R \subseteq A \times B$. The domain of the relation is A and the co-domain of the relation is B. The notation aRb signifies that $(a, b) \in R$.

A binary relation $R(A, A)$ is a relation between A and A (or a relation on A). This type of relation may always be composed with itself, and its inverse is also a binary relation on A. The identity relation on A is defined by $a\ i_A a$ for all $a \in A$.

Example 4.4 There are many examples of relations:

(i)   The relation on a set of students in a class where $(a, b) \in R$ if the height of a is greater than the height of b.

(ii) The relation between A and B where $A = \{0, 1, 2\}$ and $B = \{3, 4, 5\}$ with R given by

$$R = \{(0,3), (0,4), (1,4)\}$$

(iii) The relation less than (<) between and $\mathbb{R}$ and $\mathbb{R}$ is given by

$$\{(x,y) \in \mathbb{R}^2 : x<y\}.$$

(iv) A bank may represent the relationship between the set of accounts and the set of customers by a relation. The implementation of a bank account may be a positive integer with at most eight decimal digits.

The relationship between accounts and customers may be done with a relation $R \subseteq A \times B$, with the set A chosen to be the set of natural numbers, and the set B chosen to be the set of all human beings alive or dead. The set A could also be chosen to be $A = \{n \in \mathbb{N} : n < 10^8\}$.

A relation $R(A, B)$ may be represented pictorially. This is referred to as the graph of the relation, and it is illustrated in the diagram below. An arrow from x to y is drawn if (x, y) is in the relation. Thus, for the height relation R given by $\{(a, p), (a, r), (b, q)\}$, an arrow is drawn from a to p, from a to r and from b to q to indicate that (a, p), (a, r) and (b, q) are in the relation R.

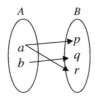

The pictorial representation of the relation makes it easy to see that the height of a is greater than the height of p and r, and that the height of b is greater than the height of q.

An n-ary relation R $(A_1, A_2, \ldots A_n)$ is a subset of $(A_1 \times A_2 \times \cdots \times A_n)$. However, an n-ary relation may also be regarded as a binary relation $R(A, B)$ with $A = A_1 \times A_2 \times \cdots \times A_{n-1}$ and $B = A_n$.

4.3.1 Reflexive, Symmetric and Transitive Relations

A binary relation on A may have additional properties such as being reflexive, symmetric or transitive. These properties are defined as

(i)   A relation on a set A is *reflexive* if $(a, a) \in R$ for all $a \in A$.
(ii) A relation R is *symmetric* if whenever $(a, b) \in R$ then $(b, a) \in R$.
(iii) A relation is *transitive* if whenever $(a, b) \in R$ and $(b, c) \in R$ then $(a, c) \in R$.

A relation that is reflexive, symmetric and transitive is termed an *equivalence relation*.

Example 4.5 (Reflexive Relation) A relation is reflexive if each element possesses an edge looping around on itself. The relation in Fig. 4.2 is reflexive.

Example 4.6 (Symmetric Relation) The graph of a symmetric relation will show for every arrow from a to b an opposite arrow from b to a. The relation in Fig. 4.3 is symmetric: i.e. whenever $(a, b) \in R$ then $(b, a) \in R$.

Example 4.7 (Transitive Relation) The graph of a transitive relation will show that whenever there is an arrow from a to b and an arrow from b to c that there is an arrow from a to c. The relation in Fig. 4.4 is transitive: i.e. whenever $(a, b) \in R$ and $(b, c) \in R$ then $(a, c) \in R$.

Example 4.8 (Equivalence Relation) The relation on the set of integers $\mathbb{Z}$ defined by $(a, b) \in R$ if $a - b = 2k$ for some $k \in \mathbb{Z}$ is an equivalence relation, and it partitions the set of integers into two equivalence classes: i.e. the even and odd integers.

Domain and Range of Relation
The domain of a relation R (A, B) is given by $\{a \in A \mid \exists b \in B \text{ and } (a, b) \in R\}$. It is denoted by **dom** R. The domain of the relation R $= \{(a, p), (a, r), (b, q)\}$ is $\{a, b\}$.

Fig. 4.2 Reflexive relation

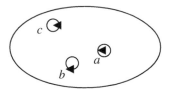

Fig. 4.3 Symmetric relation

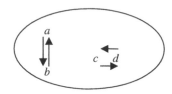

Fig. 4.4 Transitive relation

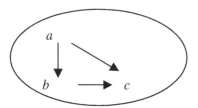

The range of a relation $R\,(A,\,B)$ is given by $\{b \in B \mid \exists a \in A$ and $(a,\,b) \in R\}$. It is denoted by **rng** R. The range of the relation $R = \{(a,\,p),\,(a,\,r),\,(b,\,q)\}$ is $\{p,\,q,\,r\}$.

Inverse of a Relation

Suppose $R \subseteq A \times B$ is a relation between A and B then the inverse relation $R^{-1} \subseteq B \times A$ is defined as the relation between B and A and is given by

$$b\,R^{-1}\,a\;\text{if and only if}\;a\,R\,b$$

That is,

$$R^{-1} = \{(b,a) \in B \times A : (a,b) \in R\}$$

Example 4.9 Let R be the relation between $\mathbb{Z}$ and $\mathbb{Z}^+$ defined by $m R n$ if and only if $m^2 = n$. Then $R = \{(m,\,n) \in \mathbb{Z} \times \mathbb{Z}^+ : m^2 = n\}$ and $R^{-1} = \{(n,\,m) \in \mathbb{Z}^+ \times \mathbb{Z} : m^2 = n\}$.

For example, $-3\;R\;9$, $-4\;R\;16$, $0\;R\;0$, $16\;R^{-1}\;-4$, $9\;R^{-1}\;-3$, etc.

Partitions and Equivalence Relations

An equivalence relation on A leads to a partition of A, and vice versa for every partition of A there is a corresponding equivalence relation.

Let A be a finite set and let $A_1, A_2, \ldots, A_n$ be subsets of A such $A_i \neq \varnothing$ for all i, $A_i \cap A_j = \varnothing$ if $i \neq j$ and $A = \bigcup_i^n A_i = A_1 \cup A_2 \cup \cdots \cup A_n$.

The sets A_i partition the set A, and these sets are called the classes of the partition (Fig. 4.5).

Theorem 4.2 (Equivalence Relation and Partitions)

An equivalence relation on A gives rise to a partition of A where the equivalence classes are given by Class $(a) = \{x \mid x \in A$ and $(a,\,x) \in R\}$. Similarly, a partition gives rise to an equivalence relation R, where $(a,\,b) \in R$ if and only if a and b are in the same partition.

Fig. 4.5 Partitions of A

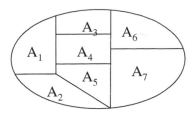

Proof Clearly, $a \in \text{Class}(a)$ since R is reflexive and clearly the union of the equivalence classes is A. Next, we show that two equivalence classes are either equal or disjoint.

Suppose $\text{Class}(a) \cap \text{Class}(b) \neq \varnothing$. Let $x \in \text{Class}(a) \cap \text{Class}(b)$ and so (a, x) and $(b, x) \in R$. By the symmetric property $(x, b) \in R$ and since R is transitive from (a, x) and (x, b) in R we deduce that $(a, b) \in R$. Therefore, $b \in \text{Class}(a)$. Suppose y is an arbitrary member of Class (b) then $(b, y) \in R$ therefore from (a, b) and (b, y) in R we deduce that (a, y) is in R. Therefore, since y was an arbitrary member of $\text{Class}(a)$ we deduce that $\text{Class}(b) \subseteq \text{Class}(a)$. Similarly, $\text{Class}(a) \subseteq \text{Class}(b)$ and so $\text{Class}(a) = \text{Class}(b)$.

This proves the first part of the theorem and for the second part, we define a relation R such that $(a, b) \in R$ if a and b are in the same partition. It is clear that this is an equivalence relation.

4.3.2   Composition of Relations

The composition of two relations $R_1(A, B)$ and $R_2(B, C)$ is given by $R_2 \circ R_1$ where $(a, c) \in R_2 \circ R_1$ if and only there exists $b \in B$ such that $(a, b) \in R_1$ and $(b, c) \in R_2$. The composition of relations is associative: i.e.

$$(R_3 \circ R_2) \circ R_1 = R_3 \circ (R_2 \circ R_1)$$

Example 4.10 (Composition of Relations) Consider a library that maintains two files. The first file maintains the serial number s of each book as well as the details of the author a of the book. This may be represented by the relation $R_1 = sR_1a$. The second file maintains the library card number c of its borrowers and the serial number s of any books that they have borrowed. This may be represented by the relation $R_2 = c \, R_2 s$.

The library wishes to issue a reminder to its borrowers of the authors of all books currently on loan to them. This may be determined by the composition of $R_1 \circ R_2$: i.e. $c \, R_1 \circ R_2 \, a$ if there is book with serial number s such that $c \, R_2 \, s$ and $s \, R_1 \, a$.

Example 4.11 (Composition of Relations) Consider sets $A = \{a, b, c\}$, $B = \{d, e, f\}$, $C = \{g, h, i\}$ and relations $R(A, B) = \{(a, d), (a, f), (b, d), (c, e)\}$ and

Fig. 4.6 Composition of
relations S o R

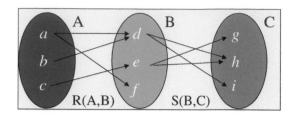

$S(B, C) = \{(d, h), (d, i), (e, g), (e, h)\}$. Then, we graph these relations and show how
to determine the composition pictorially.

S o R is determined by choosing $x \in A$ and $y \in C$ and checking if there is a route
from x to y in the graph (Fig. 4.6). If so, we join x to y in S o R. For example, if we
consider a and h we see that there is a path from a to d and from d to h, and
therefore, (a, h) is in the composition of S and R.

The union of two relations $R_1(A, B)$ and $R_2(A, B)$ is meaningful (as these are both
subsets of $A \times B$). The union $R_1 \cup R_2$ is defined as $(a, b) \in R_1 \cup R_2$ if and only
if $(a, b) \in R_1$ or $(a, b) \in R_2$.

Similarly, the intersection of R_1 and R_2 ($R_1 \cap R_2$) is meaningful and is defined
as $(a, b) \in R_1 \cap R_2$ if and only if $(a, b) \in R_1$ and $(a, b) \in R_2$. The relation R_1 is a
subset of R_2 ($R_1 \subseteq R_2$) if whenever $(a, b) \in R_1$ then $(a, b) \in R_2$.

The inverse of the relation R was discussed earlier and is given by the relation
R^{-1} where $R^{-1} = \{(b, a) \mid (a, b) \in R\}$.

The composition of R and R^{-1} yields: R^{-1} o $R = \{(a, a) \mid a \in \text{dom } R\} = i_A$ and
R o $R^{-1} = \{(b, b) \mid b \in \textbf{dom } R^{-1}\} = i_B$.

4.3.3 Binary Relations

A binary relation R on A is a relation between A and A, and a binary relation can
always be composed with itself. Its inverse is a binary relation on the same set. The
following are all relations on A:

$R^2 = R$ o R.
$R^3 = (R$ o $R)$ o R.
$R^0 = i_A$ (identity relation).
$R^{-2} = R^{-1}$ o R^{-1}.

Example 4.12 Let R be the binary relation on the set of all people P such that $(a,$
$b) \in R$ if a is a parent of b. Then, the relation R^n is interpreted as follows:

R is the parent relationship.
R^2 is the grandparent relationship.
R^3 is the great-grandparent relationship.
R^{-1} is the child relationship.

R^{-2} is the grandchild relationship.
R^{-3} is the great-grandchild relationship.

This can be generalised to a relation R^n on A where $R^n = R \circ R \circ \cdots \circ R$ (n-times). The transitive closure of the relation R on A is given by

$$R^* = \cup_{i=0}^{\infty} R^i = R^0 \cup R^1 \cup R^2 \cup \cdots R^n \cup \cdots.$$

where R^0 is the reflexive relation containing only each element in the domain of R: i.e. $R^0 = i_A = \{(a, a) \mid a \in \mathbf{dom}\ R\}$.

The positive transitive closure is similar to the transitive closure except that it does not contain R^0. It is given by

$$R^+ = \cup_{i=1}^{\infty} R^i = R^1 \cup R^2 \cup \ldots \cup R^n \cup ..$$

$a\ R^+\ b$ if and only if $a\ Rn\ b$ for some $n > 0$: i.e. there exists $c_1, c_2 \ldots c_n \in A$ such that

$$aRc_1, c_1Rc_2, \ldots, c_nRb$$

Parnas[7] introduced the concept of the limited domain relation (LD relation), and an LD relation L consists of an ordered pair (R_L, C_L) where R_L is a relation and C_L is a subset of Dom R_L. The relation R_L is on a set U and C_L is termed the competence set of the LD relation L.

The importance of LD relations is that they may be used to describe program execution. The relation component of the LD relation L describes a set of states such that if execution starts in state x it may terminate in state y. The set U is the set of states. The competence set of L is such that if execution starts in a state that is in the competence set then it is guaranteed to terminate. For a more detailed description of LD relations and their properties, see Chap. 2 of Parnas (2001).

4.3.4 Applications of Relations to Databases

A relational database management system (RDBMS) is a system that manages data using the relational model, and a relation is defined as a set of tuples that is usually represented by a table. A table is data organised in rows and columns, with the data in each column of the table of the same data type. Constraints may be employed to provide restrictions on the kinds of data that may be stored in the relations, and these Boolean expressions are a way of implementing business rules in the database.

---

[7]Parnas made important contributions to software engineering in the 1970s. He invented information hiding which is used in object-oriented design.

Relations have one or more keys associated with them, and the *key uniquely identifies the row of the table*. An index is a way of providing fast access to the data in a relational database, as it allows the tuple in a relation to be looked up directly (using the index) rather than checking all tuples in the relation.

The concept of a relational database was first described in a paper '*A Relational Model of Data for Large Shared Data Banks*' by Codd (1970). A relational database is a database that conforms to the relational model, and it may be defined as a set of relations (or tables).

Codd (Fig. 4.7) developed the *relational database model* in the late 1960s, and today, this is the standard way that information is organised and retrieved from computers. Relational databases are at the heart of systems from hospitals' patient records to airline flight and schedule information.

An *n*-ary relation R $(A_1, A_2, \ldots A_n)$ is a subset of the Cartesian product of the *n* sets: i.e. a subset of $(A_1 \times A_2 \times \cdots \times A_n)$. However, an *n*-ary relation may also be regarded as a binary relation $R(A, B)$ with $A = A_1 \times A_2 \times \cdots \times A_{n-1}$ and $B = A_n$.

The data in the relational model is defined as a set of *n*-tuples and is usually represented by a table. A table is a visual representation of the relation, and the data is organised in rows and columns.

The basic relational building block is the domain or data type (often called just type). Each row of the table represents one *n*-tuple (one tuple) of the relation, and the number of tuples in the relation is the cardinality of the relation. Consider the PART relation taken from Date (1981), where this relation consists of a heading and the body. There are five data types representing part numbers, part names, part colours, part weights and locations where the parts are stored. The body consists of a set of *n*-tuples, and the PART relation in Fig. 4.8 is of cardinality six.

There is more detailed information on the relational model and databases in Chap. 11.

Fig. 4.7 Edgar Codd

P#	PName	Colour	Weight	City
P1	Nut	Red	12	London
P2	Bolt	Green	17	Paris
P3	Screw	Blue	17	Rome
P4	Screw	Red	14	London
P5	Cam	Blue	12	Paris
P6	Cog	Red	19	London

Fig. 4.8 PART relation

4.4   Functions

A function $f : A \rightarrow B$ is a special relation such that for each element $a \in A$ there is exactly (or at most)[8] one element $b \in B$. This is written as $f(a) = b$.

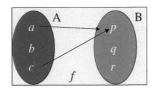

A function is a relation but not every relation is a function. For example, the relation in the diagram below is not a function since there are two arrows from the element $a \in A$.

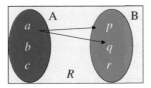

The domain of the function (denoted by **dom** f) is the set of values in A for which the function is defined. The domain of the function is A if f is a total function. The co-domain of the function is B. The range of the function (denoted **rng** f) is a subset of the co-domain and consists of:

$$\mathbf{rng}\, f = \{r | r \in B \text{ such that } f(a) = r \text{ for some } a \in A\}.$$

Functions may be partial or total. A *partial function* (or partial mapping) may be undefined for some values of A, and partial functions arise regularly in the computing field (Fig. 4.9). *Total functions* are defined for every value in A, and many functions encountered in mathematics are total.

---

[8]We distinguish between total and partial functions. A total function is defined for all elements in the domain whereas a partial function may be undefined for one or more elements in the domain.

Fig. 4.9 Domain and range
of a partial function

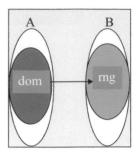

Example 4.13 (Functions) Functions are an essential part of mathematics and computer science, and there are many well-known functions such as the trigonometric functions $\sin(x)$, $\cos(x)$ and $\tan(x)$; the logarithmic function $\ln(x)$; the exponential functions e^x; and polynomial functions.

(i) Consider the partial function $f : \mathbb{R} \to \mathbb{R}$ $f(x) = 1/x$ (where $x \neq 0$).
     Then, this partial function is defined everywhere except for $x = 0$.
(ii) Consider the function $f : \mathbb{R} \to \mathbb{R}$ where

$$f(x) = x^2$$

Then, this function is defined for all $x \in \mathbb{R}$.

Partial functions often arise in computing as a program may be undefined or fail to terminate for several values of its arguments (e.g. infinite loops). Care is required to ensure that the partial function is defined for the argument to which it is to be applied.

Consider a program P that has one natural number as its input and which fails to terminate for some input values. It prints a single real result and halts if it terminates. Then, P can be regarded as a partial mapping from $\mathbb{N}$ to $\mathbb{R}$.

$$P : \mathbb{N} \to \mathbb{R}$$

Example 4.14 How many total functions $f : A \to B$ are there from A to B (where A and B are finite sets)?

Each element of A maps to any element of B, i.e. there are $|B|$ choices for each element $a \in A$. Since there are $|A|$ elements in A, the number of total functions is given by

$$|B| \, |B| . \,. ... |B| \quad (|A| \text{ times})$$
$$= |B|^{|A|} \qquad \text{total functions between } A \text{ and } B$$

Example 4.15 How many partial functions $f : A \to B$ are there from A to B (where A and B are finite sets)?

Each element of A may map to any element of B or to no element of B (as it may be undefined for that element of A). In other words, there are $|B| + 1$ choices for each element of A. As there are $|A|$ elements in A, the number of distinct partial functions between A and B is given by

$$(|B|+1)\,(|B|+1)\ldots(|B|+1) \quad (|A| \text{ times})$$
$$= (|B|+1)^{|A|}$$

Two partial functions f and g are equal if

1. dom f = dom g.
2. $f(a) = g(a)$ for all $a \in$ dom f.

A function f is less defined than a function g ($f \subseteq g$) if the domain of f is a subset of the domain of g, and the functions agree for every value on the domain of f

1. dom $f \subseteq$ dom g.
2. $f(a) = g(a)$ for all $a \in$ dom f.

The composition of functions is similar to the composition of relations. Suppose $f : A \rightarrow B$ and $g : B \rightarrow C$ then $g \circ : A \rightarrow C$ is a function, and it is written as $g \circ f(x)$ or $g(f(x))$ for $x \in A$.

The composition of functions is not commutative and this can be seen by an example. Consider the function $f : \mathbb{R} \rightarrow \mathbb{R}$ such that $f(x) = x^2$ and the function $g : \mathbb{R} \rightarrow \mathbb{R}$ such that $g(x) = x + 2$. Then

$$g \circ f(x) = g(x^2) = x^2 + 2$$
$$f \circ g(x) = f(x+2) = (x+2)^2 = x^2 + 4x + 4$$

Clearly, $g \circ f(x) \neq f \circ g(x)$ and so composition of functions is not commutative. The composition of functions is associative, as the composition of relations is associative and every function is a relation. For $f : A \rightarrow B$, $g : B \rightarrow C$, and $h : C \rightarrow D$, we have

$$h \circ (g \circ f) = (h \circ g) \circ f$$

A function $f : A \rightarrow B$ is *injective* (*one to one*) if

$$f(a_1) = f(a_2) \Rightarrow a_1 = a_2$$

For example, consider the function $f : \mathbb{R} \rightarrow \mathbb{R}$ with $f(x) = x^2$. Then, $f(3) = f(-3) = 9$ and so this function is not one to one.

A function $f : A \rightarrow B$ is *surjective* (*onto*) if given any $b \in B$ there exists an $a \in A$ such that $f(a) = b$ (Fig. 4.10). Consider the function $f : \mathbb{R} \rightarrow \mathbb{R}$ with $f(x) = x + 1$. Clearly, given any $r \in \mathbb{R}$ then $f(r - 1) = r$ and so f is onto.

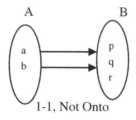

Fig. 4.10 Injective and surjective functions

Fig. 4.11 Bijective function
(one to one and onto)

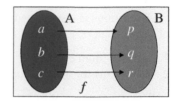

A function is *bijective* if it is one to one and onto (Fig. 4.11). That is, there is a one-to-one correspondence between the elements in A and B, and for each $b \in B$ there is a unique $a \in A$ such that $f(a) = b$.

The inverse of a relation was discussed earlier, and the relational inverse of a function $f: A \rightarrow B$ clearly exists. The relational inverse of the function may or may not be a function.

However, if the relational inverse is a function it is denoted by $f^{-1} : B \rightarrow A$. A total function has an inverse if and only if it is bijective whereas a partial function has an inverse if and only if it is injective.

The identity function $1_A : A \rightarrow A$ is a function such that $1_A(a) = a$ for all $a \in A$. Clearly, when the inverse of the function exists, then we have that $f^{-1} \, o \, f = 1_A$ and $f \, o \, f^{-1} = 1_B$.

Theorem 4.3 (Inverse of Function)

A total function has an inverse if and only if it is bijective.

Proof Suppose $f: A \rightarrow B$ has an inverse f^{-1}. Then, we show that f is bijective.

We first show that f is one to one. Suppose $f(x_1) = f(x_2)$ then

$$f^{-1}(f(x_1)) = f^{-1}(f(x_2))$$
$$\Rightarrow f^{-1} o f(x_1) = f^{-1} o f(x_2)$$
$$\Rightarrow 1_A(x_1) = 1_A(x_2)$$
$$\Rightarrow x_1 = x_2$$

Next, we first show that f is onto. Let $b \in B$ and let $a = f^{-1}(b)$ then

$$f(a) = f\left(f^{-1}(b)\right) = 1_B(b) = b \text{ and so } f \text{ is surjective}$$

The second part of the proof is concerned with showing that if $f \colon A \to B$ is bijective then it has an inverse f^{-1}. Clearly, since f is bijective, we have that for each $a \in A$, there exists a unique $b \in B$ such that $f(a) = b$.

Define $g \colon B \to A$ by letting $g(b)$ be the unique a in A such that $f(a) = b$. Then, we have that

$$g \, o f(a) = g(b) = a \text{ and } f \, o \, g(b) = f(a) = b.$$

Therefore, g is the inverse of f.

4.4.1 Application of Functions to Functional Programming

Functional programming involves the evaluation of mathematical functions, whereas imperative programming involves the execution of sequential (or iterative) commands that change the state. For example, the assignment statement alters the value of a variable, and the value of a given variable x may change during program execution.

There are no changes of state for functional programs, and the fact that the value of x will always be the same makes it easier to reason about functional programs than imperative programs. Functional programming languages provide *referential transparency* : i.e. equals may be substituted for equals, and if two expressions have equal values, then one can be substituted for the other in any larger expression without affecting the result of the computation.

Functional programming languages use higher order functions,[9] recursion, lazy and eager evaluation, monads,[10] and Hindley–Milner type inference systems.[11] These languages are mainly used in academia, but there has been some industrial use, including the use of Erlang for concurrent applications in industry. Alonzo Church developed Lambda Calculus in the 1930s, and it provides an abstract framework for describing mathematical functions and their evaluation. It provides the foundation for functional programming languages. Church employed lambda Calculus to prove that there is no solution to the decision problem for first-order arithmetic (O'Regan 2017).

---

[9]Higher order functions are functions that take functions as arguments or return a function as a result. They are known as operators (or functionals) in mathematics, and one example is the derivative function $\frac{dy}{dx}$ that takes a function as an argument and returns a function as a result.

[10]Monads are used in functional programming to express input and output operations without introducing side effects. The Haskell functional programming language makes use of this feature.

[11]This is the most common algorithm used to perform type inference. Type inference is concerned with determining the type of the value derived from the eventual evaluation of an expression.

The original Calculus developed by Church was untyped, but typed lambda calculi have since been developed. Any computable function can be expressed and evaluated using lambda Calculus, but there is no general algorithm to determine whether two arbitrary lambda Calculus expressions are equivalent. Lambda Calculus influenced functional programming languages such as Lisp, ML and Haskell.

Functional programming uses higher order functions which take functions as arguments and return functions as results. The derivative function $^d/_{dx} f(x) = f'(x)$ is a higher order function, which takes a function as an argument and returns a function as a result. Higher order functions may employ currying (a technique developed by Schönfinkel) which allows a function with several arguments to be applied to each of its arguments one at a time, with each application returning a new (higher order) function that accepts the next argument. This allows a function of n-arguments to be treated as n applications of a function with 1-argument.

John McCarthy developed LISP at MIT in the late 1950s, and this language includes many of the features found in modern functional programming languages.[12] Robin Milner designed the ML programming language in the early 1970s. David Turner developed Miranda in the mid-1980s. The Haskell programming language was released in the late 1980s. There is more detailed information on functional programming in Chap. 8.

4.5 Number Theory

Number theory is the branch of mathematics that is concerned with the mathematical properties of the natural numbers and integers. These include properties such as the parity of a number, divisibility, additive and multiplicative properties, whether a number is prime or composite, the prime factors of a number, the greatest common divisor and least common multiple of two numbers, and so on.

Number theory has many applications in computing including cryptography and coding theory. For example, the RSA public key cryptographic system relies on its security due to the infeasibility of the integer factorization problem for large numbers.

There are several unsolved problems in number theory and especially in prime number theory. For example, Goldbach's[13] Conjecture states that every even integer greater than two is the sum of two primes, and this result has not been proved to date. Fermat's[14] last theorem (Fig. 4.12) states that there is no integer

---

[12]Lisp is a multi-paradigm language rather than a functional programming language.

[13]Goldbach was an eighteenth-century German mathematician and Goldbach's conjecture has been verified to be true for all integers $n < 12 \times 10^{17}$.

[14]Pierre de Fermat was a seventeenth-century French civil servant and amateur mathematician. He occasionally wrote to contemporary mathematicians announcing his latest theorem without providing the accompanying proof and inviting them to find the proof. The fact that he never revealed his proofs caused a lot of frustration among his contemporaries, and in his announcement of his

Fig. 4.12 Pierre de Fermat

solution to $x^n + y^n = z^n$ for $n > 2$, and this result remained unproved for over three hundred years until Andrew Wiles finally proved it in the mid-1990s.

The natural numbers $\mathbb{N}$ consist of the numbers $\{1, 2, 3, \ldots\}$. The integer numbers $\mathbb{Z}$ consist of $\{\ldots, -2, -1, 0, 1, 2, \ldots\}$. The rational numbers $\mathbb{Q}$ consist of all numbers of the form $\{^p/_q$ where p and q are integers and $q \neq 0\}$. The real numbers $\mathbb{R}$ is defined to be the set of converging sequences of rational numbers and they are a superset of the rational numbers. They contain the rational and irrational numbers. The complex numbers $\mathbb{C}$ consist of all numbers of the form $\{a + bi$ where $a, b \in \mathbb{R}$ and $i = \sqrt{-1}\}$.

Pythagorean triples are combinations of three whole numbers that satisfy Pythagoras' equation $x^2 + y^2 = z^2$. There are an infinite number of such triples, and an example of such a triple is 3, 4, 5 since $3^2 + 4^2 = 5^2$.

The Pythagoreans discovered the mathematical relationship between the harmony of music and numbers, and their philosophy was that numbers are hidden in everything from music to science and nature. This led to their philosophy that 'everything is number'.

4.5.1 Elementary Number Theory

A square number is an integer that is the square of another integer. For example, the number 4 is a square number since $4 = 2^2$. Similarly, the number 9 and the number

---

famous last theorem, he stated that he had a wonderful proof that was too large to include in the margin. He corresponded with Pascal and they did some early work on the mathematical rules of games of chance and early probability theory. He also did some early work on the Calculus.

16 are square numbers. A number n is a square number if and only if one can arrange the n points in a square.

The square of an odd number is odd whereas the square of an even number is even. This is clear since an even number is of the form $n = 2k$ for some k, and so $n^2 = 4k^2$ which is even. Similarly, an odd number is of the form $n = 2k + 1$ and so $n^2 = 4k^2 + 4k + 1$ which is odd.

A rectangular number n may be represented by a vertical and horizontal rectangle of n points. For example, the number 6 may be represented by a rectangle with length 3 and breadth 2, or a rectangle with length 2 and breadth 3. Similarly, the number 12 can be represented by a 4×3 or a 3×4 rectangle.

A triangular number (Fig. 4.13) n may be represented by an equilateral triangle of n points. It is the sum of k natural numbers from 1 to k. = That is,

$$n = 1 + 2 + \cdots + k$$

Parity of Integers

The parity of an integer refers to whether the integer is odd or even. An integer n is odd if there is a remainder of one when it is divided by two, and it is of the form $n = 2k + 1$. Otherwise, the number is even and of the form $n = 2k$.

The sum of two numbers is even if both are even or both are odd. The product of two numbers is even if at least one of the numbers is even. These properties are expressed as follows:

Even $\pm$ even = even,
Even $\pm$ odd = odd,
Odd $\pm$ odd = even,
Even $\times$ even = even,
Even $\times$ odd = even and
Odd $\times$ odd = odd.

Divisors

Let a and b be integers with $a \neq 0$ then a is said to be a divisor of b (denoted by $a|b$) if there exists an integer k such that $b = ka$.

A divisor of n is called a *trivial divisor* if it is either 1 or n itself; otherwise, it is called a *non-trivial divisor*. A *proper divisor* of n is a divisor of n other than n itself.

Fig. 4.13 Triangular
numbers

Fig. 4.14 Marin Mersenne

Definition

(*Prime Number*) A *prime number* is a number (>1) whose only divisors are trivial. There are an infinite number of prime numbers.

The *fundamental theorem of arithmetic* states that every integer number can be factored as the product of prime numbers.

Mersenne Primes

Mersenne primes are prime numbers of the form $2^p - 1$ where p is a prime. They are named after Marin Mersenne (Fig. 4.14) who was a seventeenth-century French monk. Mersenne did some early work in identifying primes of this format, and there are 47 known Mersenne primes. It is an open question as to whether there are an infinite number of Mersenne primes.

For more information on number theory, see O'Regan (2017).

4.6 Automata Theory

Automata theory is the branch of computer science that is concerned with the study of abstract machines and automata. These include finite-state machines, pushdown automata and Turing machines. Finite-state machines are abstract machines that may be in one of a finite number of states. These machines are in only one state at a time (current state), and the input symbol causes a transition from the current state to the next state. Finite-state machines have limited computational power due to memory and state constraints, but they have been applied to several fields including communication protocols, neurological systems and linguistics.

Pushdown automata have greater computational power, and they contain extra memory in the form of a stack from which symbols may be pushed or popped. The state transition is determined from the current state of the machine, the input symbol and the element on the top of the stack. The action may be to change the state and/ or push/pop an element from the stack.

The Turing machine is the most powerful model for computation, and this theoretical machine is equivalent to an actual computer in the sense that it can compute the same set of functions. The memory of the Turing machine is a tape that consists of a potentially infinite number of one-dimensional cells. The Turing machine provides a mathematical abstraction of computer execution and storage, as well as providing a mathematical definition of an algorithm. However, Turing machines are not suitable for programming, and therefore, they do not provide a good basis for studying programming and programming languages.

4.6.1   Finite-State Machines

Warren McCulloch and Walter Pitts published early work on finite-state automata in 1943. They were interested in modelling the thought process for humans and machines. Moore and Mealy developed this work further, and their finite-state machines are referred to as the '*Mealy machine*' and the '*Moore machine*'. The Mealy machine determines its outputs through the current state and the input, whereas the output of Moore's machine is based upon the current state alone.

Definition
(*Finite-State Machine*) A finite-state machine (FSM) is an abstract mathematical machine that consists of a finite number of states. It includes a start state q_0 in which the machine is in initially, a finite set of states Q, an input alphabet Σ, a state transition function δ and a set of final accepting states F (where $F \subseteq Q$).
    The state transition function δ takes the current state and an input symbol, and returns the next state. That is, the transition function is of the form

$$\delta : Q \times \Sigma \rightarrow Q$$

The transition function provides rules that define the action of the machine for each input symbol, and its definition may be extended to provide output as well as a transition of the state. State diagrams are used to represent finite-state machines, and each state accepts a finite number of inputs. A finite-state machine may be deterministic or non-deterministic. A *deterministic machine* changes to exactly (or at most)[15] one state for each input transition, whereas a *non-deterministic machine* may have a choice of states to move to for a particular input symbol.
    Finite-state automata can compute only very primitive functions, and so they are not adequate as a model for computing. There are more powerful automata such as the Turing machine that is essentially a finite automaton with a potentially infinite storage (memory). Anything that is computable is computable by a Turing machine.

---

[15]The transition function may be undefined for a particular input symbol and state.

Fig. 4.15 Deterministic FSM

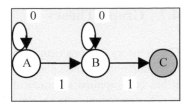

Table 4.2 Properties of set operations

State	0	1
A	A	B
B	B	C
C	–	–

A finite-state machine can model a system that has a finite number of states, and a finite number of inputs/events that can trigger transitions between states. The behaviour of the system at a point in time is determined from the current state and input, with behaviour defined for the possible input to that state. The system starts in an initial state.

A finite-state machine (also known as finite-state automata) is defined mathematically as a quintuple $(\Sigma, Q, \delta, q_0, F)$. The alphabet of the FSM is given by Σ; the set of states is given by Q; the transition function is defined by $\delta : Q \times \Sigma \rightarrow Q$; the initial state is given by q_0; and the set of accepting states is given by F (where F is a subset of Q). A string is given by a sequence of alphabet symbols: i.e. $s \in \Sigma^*$, and the transition function δ may be extended to $\delta^* : Q \times \Sigma^* \rightarrow Q$.

A string $s \in \Sigma^*$ is accepted by the finite-state machine if $\delta^*(q_0, s) = q_f$ where $q_f \in F$, and the set of all strings accepted by a finite-state machine is termed the *language* generated by the machine. A finite-state machine is termed *deterministic* (Fig. 4.15) if the transition function δ is a function,[16] and otherwise (where it is a relation) it is said to be *non-deterministic*. A non-deterministic automaton is one for which the next state is not uniquely determined from the present state and input symbol, and the transition may be to a set of states rather than a single state.

For the example above, the input alphabet is given by $\Sigma = \{0, 1\}$, the set of states by {A, B, C}, the start state by A and the final state by {C}, and the transition function is given by the state transition table (Table 4.2). The language accepted by the automata is the set of all binary strings that end with a one that contains exactly two ones.

For more detailed information on automata theory, see O'Regan (2017).

---

[16]It may be a total or a partial function.

4.7 Graph Theory

Graph theory is a practical branch of mathematics that deals with the arrangements of certain objects known as vertices (or nodes) and the relationships between them. It has been applied to practical problems such as the modelling of computer networks, determining the shortest driving route between two cities, the link structure of a website, the travelling salesman problem and the four-colour problem.[17]

The map of the London underground does not represent every feature of the city of London, as it includes only material that is relevant to the users of the London underground transport system. The exact geographical location of the stations is unimportant, as the essential information is how the stations are interconnected to one another. This allows a passenger to plan a route from one station to another. That is, the map of the London underground is essentially a model of the transport system that shows how the stations are interconnected.

The seven bridges of Königsberg in Prussia[18] (Fig. 4.16) is one of the earliest problems in graph theory. The city was set on both sides of the Pregel River in the early eighteenth century, and it consisted of two large islands that were connected to each other and the mainland by seven bridges. The problem was to find a walk through the city that would cross each bridge once and once only.

Euler showed that the problem had no solution, and his analysis helped to lay the foundations for graph theory as a discipline. This problem is concerned with the question as to whether it is possible to travel along the edges of a graph starting from a vertex and returning to it and travelling along each edge exactly once. An Euler path in a graph G is a simple path containing every edge of G.

Euler noted, in effect, that for a walk through a graph traversing each edge exactly once depends on the *degree* of the nodes (i.e. the number of edges touching it). He showed that a necessary and sufficient condition for the walk is that the graph is connected and has zero or two nodes of odd degree. For the Königsberg graph, the four nodes (i.e. the land masses) have an odd degree (Fig. 4.17).

A *graph* is a collection of objects that are interconnected in some way. The objects are typically represented by vertices (or nodes), and the interconnections between them are represented by edges (or lines). We distinguish between directed and adirected graphs, where a *directed graph* is mathematically equivalent to a binary relation, and an *adirected (undirected) graph* is equivalent to a symmetric binary relation. For more detailed information on graph theory, see O'Regan (2017).

---

[17]The four-colour theorem states that given any map it is possible to colour the regions of the map with no more than four colours such that no two adjacent regions have the same colour. This result was finally proved in the mid-1970s.

[18]Königsberg (now called Kaliningrad) was founded in the thirteenth century by Teutonic Knights and was one of the cities of the Hanseatic League. It was the historical capital of East Prussia (part of Germany), and it was annexed by Russia at the end of the Second World War. The famous German philosopher, Immanuel Kant, spent all his life in the city and is buried there.

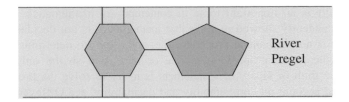

Fig. 4.16 Königsberg seven bridges problem

Fig. 4.17 Königsberg graph

4.8   Computability and Decidability

It is impossible for a human or machine to write out all members of an infinite countable set, such as the set of natural numbers $\mathbb{N}$. However, humans can do something quite useful in the case of certain enumerable infinite sets: they can give explicit instructions (that may be followed by a machine or another human) to produce the nth member of the set for an arbitrary finite n. The problem remains that for all but a finite number of values of n, it will be physically impossible for any human or machine to carry out the computation, due to the limitations on the time available for computation, the speed at which the individual steps in the computation may be carried out, and due to finite materials.

The intuitive meaning of computability is in terms of an algorithm (or effective procedure) that specifies a set of instructions to be followed to complete the task. That is, a function f is *computable* if there exists an algorithm that produces the value of f correctly for each possible argument of f. The computation of f for an argument x just involves following the instructions in the algorithm, and it produces the result $f(x)$ in a finite number of steps if x is in the domain of f. If x is not in the domain of f then the algorithm may produce an answer saying so or it might run forever never halting. A computer program implements an algorithm and algorithms are discussed in more detail in Chap. 5.

The concept of computability may be made precise in several equivalent ways such as Church's *lambda Calculus*, *recursive function theory* or by the theoretical *Turing machines*.[19] These are all equivalent and perhaps the most well known is the Turing machine (O'Regan 2017), where the set of functions that are computable are those that are computable by a Turing machine.

---

[19]The Church–Turing thesis states that anything that is computable is computable by a Turing machine.

Decidability is an important topic in contemporary mathematics. Church and Turing independently showed in 1936 that mathematics is not decidable. In other words, there is no mechanical procedure (i.e. algorithm) to determine whether an arbitrary mathematical proposition is true or false, and so the only way is to determine the truth or falsity of a statement is to try to solve the problem. For a more detailed account on computability and decidability, see O'Regan (2017).

4.9 Review Questions

1. What is a set? *A* relation? *A* function?
2. Explain the difference between a partial and a total function
3. Explain the difference between a relation and a function
4. Determine $A \times B$ where $A = \{a, b, c, d\}$ and $B = \{1, 2, 3\}$
5. Determine $A \mathrel{\Delta} B$ where $A = \{a, b, c, d\}$ and $B = \{c, d, e\}$
6. What is the graph of the relation $\leq$ on the set $A = \{2, 3, 4\}$?
7. What is the domain and range of $R = \{(a, p), (a, r), (b, q)\}$?
8. Determine the inverse relation R^{-1} where $R = \{(a, 2), (a, 5), (b, 3), (b, 4), (c, 1)\}$
9. Determine the inverse of the function $f : \mathbb{R} \times \mathbb{R} \to \mathbb{R}$ defined by $f(x) = \frac{x-2}{x-3} (x \neq 3)$ and $f(3) = 1$.
10. Give examples of injective, surjective and bijective functions
11. Explain the differences between imperative programming languages and functional programming languages.

4.10 Summary

This chapter introduced essential mathematics for computing including set theory, relations and functions. Sets are collections of well-defined objects; a relation between A and B indicates relationships between members of the sets A and B; and functions are a special type of relation where there is at most one relationship for each element $a \in A$ with an element in B.

A binary relation $R (A, B)$ is a subset of the Cartesian product $(A \times B)$ of A and B where A and B are sets. The domain of the relation is A and the co-domain of the relation is B. An n-ary relation $R (A_1, A_2, \ldots A_n)$ is a subset of $(A_1 \times A_2 \times \cdots \times A_n)$.

A total function $f : A \to B$ is a special relation such that for each element $a \in A$ there is exactly one element $b \in B$. This is written as $f(a) = b$. A function is a relation but not every relation is a function.

Functional programming is quite distinct from imperative programming in that there is no change of state, and the value of the variable x remains the same during program execution. This makes functional programs easier to reason about than imperative programs.

Automata theory is the branch of computer science that is concerned with the study of abstract machines and automata. These include finite-state machines, pushdown automata and Turing machines. Graph theory is a practical branch of mathematics that deals with the arrangements of certain objects known as vertices (or nodes) and the relationships between them.

Chapter 5
Introduction to Algorithms

5.1 Introduction

An *algorithm* is a well-defined procedure for solving a problem, and it consists of a sequence of steps that takes a set of values as input and produces a set of values as output. It is an exact specification of how to solve the problem, and it explicitly defines the procedure so that a computer program may implement the algorithm. The origin of the word '*algorithm*' is from the name of the ninth-century Persian mathematician, Mohammed Al Khwarizmi.

It is essential that the algorithm is correct, and that it terminates in a reasonable time. This may require mathematical analysis of the algorithm to demonstrate its correctness and efficiency, and to show that termination is within an acceptable timeframe. There may be several algorithms to solve a problem, and so the choice of the best algorithm (e.g. fastest/most efficient) needs to be considered. For example, there are several well-known sorting algorithms (e.g. *Merge sort* and *insertion sort*), and the Merge sort algorithm is more efficient [$o(n \lg n)$] than the insertion sort algorithm [$o(n^2)$].

An algorithm may be implemented by a computer program written in some programming language (e.g. C++ or Java). The speed of the program depends on the algorithm employed, the input value(s), how the algorithm has been implemented in the programming language, the compiler, the operating system and the computer hardware.

© Springer International Publishing AG, part of Springer Nature 2018
G. O'Regan, *World of Computing*,
https://doi.org/10.1007/978-3-319-75844-2_5

An algorithm may be described in natural language (care is needed to avoid ambiguity), but it is more common to use a more precise formalism for its description. These include *pseudocode* (an informal high-level language description), flowcharts, a programming language such as C or Java, or a formal specification language such as VDM or Z. We shall mainly use natural language and pseudocode to describe an algorithm. One of the earliest algorithms developed was Euclid's algorithm for determining the greatest common divisor of two natural numbers, and it is described in the next section.

5.2   Early Algorithms

Euclid lived in Alexandria during the early Hellenistic period,[1] and he is considered the father of geometry and the deductive method in mathematics. His systematic treatment of geometry and number theory is published in the 13 books of the Elements (Heath 1956). It starts from five axioms, five postulates and 23 definitions to logically derive a comprehensive set of theorems in geometry.

His method of proof was generally constructive, in that as well as demonstrating the truth of the theorem, a construction of the required entity was provided. He employed some indirect proofs and one example was his proof that there are an infinite number of prime numbers. The procedure is to assume the opposite of what one wishes to prove and to show that a contradiction results. This means that the original assumption must be false, and the theorem is established.

1. Suppose there are a finite number of primes (say n primes).
2. Multiply all n primes together and add 1 to form N.

$$(N = p_1 * p_2 * \ldots * p_n + 1)$$

3. N is not divisible by $p_1, p_2, \ldots, p_n$ as dividing by any of these gives a remainder of one.
4. Therefore, N must either be prime or divisible by some other prime that was not included in the original list.
5. Therefore, there must be at least $n + 1$ primes.
6. This is a contradiction (it was assumed that there are exactly n primes).
7. Therefore, the assumption that there are a finite number of primes is false.
8. Therefore, there are an infinite number of primes.

---

[1]This refers to the period following the conquests of Alexander the Great, which led to the spread of Greek culture throughout the Middle East and Egypt.

His proof that there are an infinite number of primes is indirect, and he does not present an algorithm to as such to construct the set of prime numbers. We present the well-known Sieve of Eratosthenes algorithm for determining the prime numbers up to a given number n later in the chapter.

The material in the Euclid's Elements is a systematic development of geometry starting from the small set of axioms, postulates and definitions. It leads to many well-known mathematical results such as Pythagoras's theorem, Thales theorem, sum of angles in a triangle, prime numbers, greatest common divisor and least common multiple, Euclidean algorithm, areas and volumes, tangents to a point and algebra.

5.2.1 Greatest Common Divisors (GCD)

Let a and b be integers not both zero. The *greatest common divisor d* of a and b is a divisor of a and b (i.e. $d \mid a$ and $d \mid b$), and it is the largest such divisor (i.e. if $k \mid a$ and $k \mid b$ then $k \mid d$). It is denoted by gcd (a, b).

Properties of Greatest Common Divisors

(i) Let a and b be integers not both zero, then exist integers x and y such that

$$d = \gcd(a, b) = ax + by$$

(ii) Let a and b be integers not both zero, then the set $S = \{ax + by$ where $x, y \in \mathbb{Z}\}$ is the set of all multiples of $d = $ gcd (a, b).

5.2.2 Euclid's Greatest Common Divisor Algorithm

Euclid's algorithm is one of the oldest known algorithms, and it provides the procedure for finding the greatest common divisor of two numbers a and b. It appears in Book VII of Euclid's Elements, but the algorithm was known prior to Euclid (Fig. 5.1).

The inputs for the gcd algorithm consist of two natural numbers a and b, and the output of the algorithm is d (the greatest common divisor of a and b). It is computed as follows:

$$\gcd(a, b) = \begin{cases} \text{Check if } b \text{ is zero. If so, then } a \text{ is the gcd.} \\ \text{Otherwise, the gcd } (a, b) \text{ is given by gcd } (b, a \bmod b). \end{cases}$$

Fig. 5.1 Euclid of
Alexandria

It is also possible to determine integers p and q such that $ap + bq = \gcd(a, b)$.

The (informal) proof of the Euclidean algorithm is as follows. Suppose a and b are two positive numbers whose greatest common divisor is to be determined, and let r be the remainder when a is divided by b.

1. Clearly, $a = qb + r$ where q is the quotient of the division.
2. Any common divisor of a and b is also a divider or r (since $r = a - qb$).
3. Similarly, any common divisor of b and r will also divide a.
4. Therefore, the greatest common divisor of a and b is the same as the greatest common divisor of b and r.
5. The number r is smaller than b, and we will reach $r = 0$ in finitely many steps.
6. The process continues until $r = 0$.

Comment 5.1

Algorithms are fundamental in computing as they define the procedure by which a problem is solved. A computer program implements the algorithm in some programming language.

Next, we deal with the Euclidean algorithm more formally, and we start with a basic lemma.

Lemma *Let a, b, q and r be integers with $b > 0$ and $0 \leq r < b$ such that $a = bq + r$. Then, $\gcd(a, b) = \gcd(b, r)$.*

Proof Let $K = \gcd(a, b)$ and let $L = \gcd(b, r)$ then we need to show that $K = L$. Suppose m is a divisor of a and b, then as $a = bq + r$ we have m is a divisor of r and so any common divisor of a and b is a divisor of r. Therefore, the greatest common divisor K of a and b is a divisor of r. Similarly, any common divisor n of

b and r is a divisor of a. Therefore, the greatest common divisor L of b and r is a divisor of a. That is, K divides L and L divides K and so $L = K$, and so the greatest common divisor of a and b is equal to the greatest common divisor of b and r.

Euclid's Algorithm (more formal proof)
Euclid's algorithm for finding the greatest common divisor of two positive integers a and b involves a repeated application of the division algorithm as follows:

$$a = bq_0 + r_1 \quad 0 < r_1 < b$$
$$b = r_1q_1 + r_2 \quad 0 < r_2 < r_1$$
$$r_1 = r_2q_2 + r_3 \quad 0 < r_3 < r_2$$
$$\ldots \ldots \ldots \ldots$$
$$\ldots \ldots \ldots \ldots$$
$$r_{n-2} = r_{n-1}q_{n-1} + r_n \quad 0 < r_n < r_{n-1}$$
$$r_{n-1} = r_n q_n$$

Then, r_n (i.e. the last non-zero remainder) is the greatest common divisor of a and b: i.e. $\gcd(a, b) = r_n$.

Proof It is clear from the construction that r_n is a divisor of $r_{n-1}, r_{n-2}, \ldots, r_3, r_2, r_1$ and of a and b. Clearly, any common divisor of a and b will also divide r_n. Using the results from the lemma above, we have

$$\gcd(a, b)$$
$$= \gcd(b, r_1)$$
$$= \gcd(r_1 r_2)$$
$$= \ldots$$
$$= \gcd(r_{n-2} r_{n-1})$$
$$= \gcd(r_{n-1}, r_n)$$
$$= r_n$$

5.2.3 Sieve of Eratosthenes Algorithm

Eratosthenes was a Hellenistic mathematician and scientist who worked in the famous library in Alexandria. He devised a system of latitude and longitude, and he was the first person to estimate the size of the circumference of the earth. He developed a famous algorithm (the well-known *Sieve of Eratosthenes algorithm*) for determining the prime numbers up to a given number n.

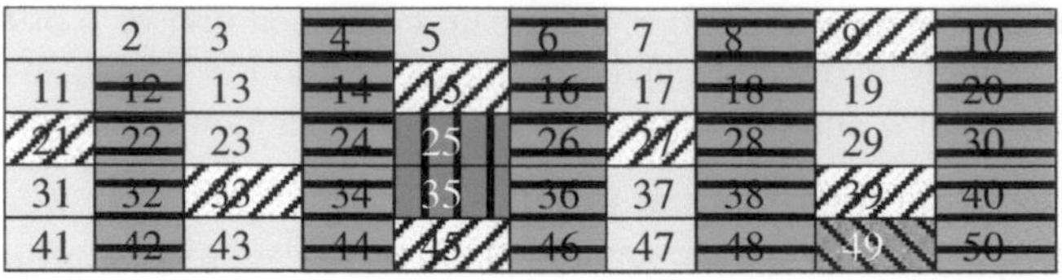

	2	3	4	5	6	7	8	9	10
11	12	13	14	15	16	17	18	19	20
21	22	23	24	25	26	27	28	29	30
31	32	33	34	35	36	37	38	39	40
41	42	43	44	45	46	47	48	49	50

Fig. 5.2 Primes between 1 and 50

The algorithm involves listing all numbers from 2 up to n. The first step is to remove all multiples of 2 up to n; the second step is to remove all multiples of 3 up to n; and so on (Fig. 5.2).

The kth step involves removing multiples of the kth prime p_k up to n and the steps in the algorithm continue while $p_k \leq \sqrt{n}$. The numbers remaining in the list are the prime numbers from 2 to n.

1. List the integers from 2 to n.
2. For each prime p_k up to $\sqrt{n}$ remove all multiples of p_k.
3. The numbers remaining are the prime numbers between 2 and n.

The list of primes between 1 and 50 are then given by 2, 3, 5, 7, 11, 13, 17, 19, 23, 29, 31, 37, 41, 43 and 47.

The steps in the algorithm may also be described as follows (in terms of two lists):

1. Write a list of the numbers from 2 to the largest number to be tested. This first list is called A.
2. A second List B is created to list the primes. It is initially empty.
3. The number 2 is the first prime number, and it is added to List B.
4. Strike off (or remove) all multiples of 2 from List A.
5. The first remaining number in List A is a prime number and this prime number is added to List B.
6. Strike off (or remove) this number and all multiples of it from List A.
7. Repeat steps 5 through 7 until no more numbers are left in List A.

5.2.4   Early Cipher Algorithms

Julius Caesar employed a *substitution cipher* on his military campaigns to ensure that important messages were communicated safely. The Caesar cipher is a very simple encryption algorithm, and it involves the substitution of each letter in the *plaintext* (i.e. the original message) by a letter a fixed number of positions down in the alphabet. The Caesar encryption algorithm involves a shift of three positions and causes the letter B to be replaced by E, the letter C by F and so on. The Caesar

Alphabet Symbol	abcde fghij klmno pqrst uvwxyz
Cipher Symbol	dfegh ijklm nopqr stuvw xyzabc

Fig. 5.3 Caesar cipher

cipher is easily broken, as the frequency distribution of letters may be employed to determine the mapping. The Caesar cipher is defined as follows (Fig. 5.3).

The process of enciphering a message (i.e. the plaintext) involves mapping each letter in the plaintext to the corresponding cipher letter. For example, the encryption of 'summer solstice' involves:

$$\text{Plaintext :} \quad \text{summer solstice}$$
$$\text{Cipher Text} \quad \text{vxpphu vrovwleh}$$

The decryption involves the reverse operation: i.e. for each cipher letter, the corresponding plaintext letter is determined from the table.

$$\text{Cipher Text} \quad \text{vxpphu vrovwleh}$$
$$\text{Plaintext :} \quad \text{summer solstice}$$

The Caesar encryption algorithm may be expressed formally using modular arithmetic. The numbers 0–25 represent the alphabet letters, and the algorithm is expressed using addition (modula 26) to yield the encrypted cipher. The encoding of the plaintext letter x is given by

$$c = x + 3 (\text{mod } 26)$$

Similarly, the decoding of a cipher letter represented by the number c is given by

$$x = c - 3 (\text{mod } 26)$$

The emperor Augustus[2] employed a similar substitution cipher (with a shift key of 1). The Caesar cipher remained in use up to the early twentieth century. However, by then frequency analysis techniques were available to break the cipher. The *Vignère cipher* uses a Caesar cipher with a different shift at each position in the text. The value of the shift to be employed with each plaintext letter is defined using a repeating keyword.

---

[2]Augustus was the first Roman emperor and his reign ushered in a period of peace and stability following the bitter civil war that occurred after the assassination of Julius Caesar. Augustus was the adopted son of Julius Caesar (he was called Octavian before he became emperor). The civil war broke out between Mark Anthony and Octavian, and Anthony and Cleopatra were defeated by Octavian and Agrippa at the battle of Actium in 31 B.C.

5.3   Sorting Algorithms

One of the most common tasks to be performed in a computer program is that of sorting (e.g. consider the problem of sorting a list of names or numbers). This has led to the development of many sorting algorithms (e.g. selection sort, bubble sort, insertion sort, Merge sort and quicksort) as sorting is a fundamental task to be performed.

For example, consider the problem of specifying the algorithm for sorting a sequence of n numbers. Then, the input to the algorithm is $\langle x_1, x_2, \ldots x_n \rangle$, and the output is $\langle x_1', x_2', \ldots x_n' \rangle$, where $x_1' \leq x_2' \leq \ldots \leq x_n'$. Further, $\langle x_1', x_2', \ldots x_n' \rangle$ is a permutation of $\langle x_1, x_2, \ldots x_n \rangle$: i.e. the same numbers are in both sequences except that the sorted sequence is in ascending order, whereas no order is imposed on the original sequence.

Insertion sort is an efficient algorithm for sorting a small list of elements. It iterates over the input sequence, examines the next input element during the iteration and builds up the sorted output sequence. During the iteration, insertion sort removes the next element from the input data, and it then finds and inserts it into the location where it belongs in the sorted list. This continues until there are no more input elements to process (Fig. 5.4).

We first give an example of insertion sort and then give a more formal definition of the algorithm. The example considered is that of the insertion sort algorithm applied to the sequence $A = \langle 5, 3\ 1, 4 \rangle$. The current input element for each iteration is highlighted, and the arrow points to the location where it is inserted in the sorted sequence. For each iteration, the elements to the left of the current element are already in increasing order, and the operation of insertion sort is to move the current element to the appropriate position in the ordered sequence.

We shall assume that we have an unsorted array A with n elements that we wish to sort. The operation of insertion sort is to rearrange the elements of A within the array, and the output is that the array A contains the sorted output sequence.

```
Insertion Sort
for i from 2 to n do
    C ← A[i]
    j ← i − 1
    while j > 0 and A[j] > C do
        A[j + 1] ← A[j]
        j ← j − 1
    A[j + 1] ← C
```

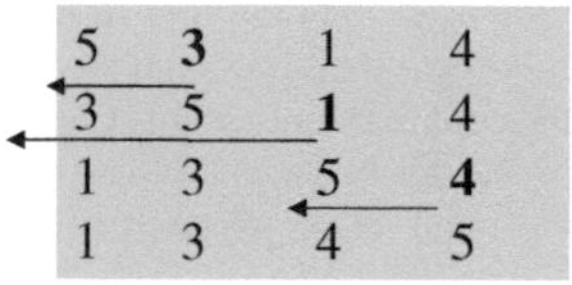

Fig. 5.4 Insertion sort example

5	3	1	4
3	5	1	4
1	3	5	4
1	3	4	5

The analysis of an algorithm involves determining its efficiency, and establishing the resources that it requires (e.g. memory and bandwidth), as well as determining the computational time required. The time taken by the insertion sort algorithm depends on the size of the input sequence (clearly a large sequence will take longer to sort than a short sequence), and on the extent to which the sequences are already sorted. The worst-case running time for the insertion sort algorithm is of order n^2— i.e. $o(n^2)$, where n is the size of the sequence to be sorted (the average case is also of order n^2 with the best case linear).

There are a number of ways to design sorting algorithms, and the insertion sort algorithm uses an incremental approach, with the sub-array $A[1 \ldots i - 1]$ already sorted and the element $A[i]$ is then inserted into its correct place to yield the sorted array $A[1 \ldots i]$.

Another approach is to employ divide and conquer techniques, and this technique is used in the Merge sort algorithm. This is a more efficient algorithm than insertion sort, and the divide and conquer method involves breaking a problem down into several sub-problems, and then solving each problem separately. The problem-solving may involve recursion or directly solving the sub-problem (if it is small enough), and then combining the solutions to the sub-problems into the solution for the original problem. The Merge sort algorithm involves three steps (divide, conquer and combine):

1. *Divide* the List A (with n elements) to be sorted into two subsequences (each with $n/2$ elements).
2. Sort each of the subsequences by calling Merge sort recursively (*Conquer*).
3. Merge the two sorted subsequences to produce a single sorted list (*Combine*).

The recursive part of the Merge sort algorithm bottoms out when the sequence to be sorted is of length 1, as for this case the sequence is of length 1 which is already (trivially) sorted. The key operation then (where all the work is done) is the combine step that Merges two sorted sequences to produce a single sorted sequence. The Merge sort algorithm may also be described as follows:

1. Divide the sequence (with n elements) to be sorted into n subsequences each with one element (a sequence with one element is sorted).
2. Repeatedly Merge subsequences to form new subsequences (each new subsequence is sorted), until there is only one remaining subsequence (the sorted sequence).

First, we consider an example (Fig. 5.5) to illustrate how the Merge sort algorithm operates, and we then give a formal definition.

It may be seen from the example that the list is repeatedly divided into equal halves with each iteration, until we get to the atomic values that can no longer be divided. The lists are then combined in the order in which they were broken down, and this involves comparing the elements of both lists and combining them to form a sorted list. The merging continues in this way until there are no more lists to

Fig. 5.5 Merge sort example

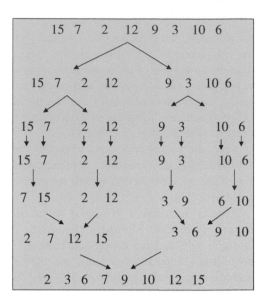

Merge, and the list remaining is the sorted list. The formal definition of Merge sort is as follows:

Merge Sort(A, m, n)
If $m < n$ then
$\quad r \leftarrow (m + n)$ div 2
$\quad$ Merge Sort(A, m, r)
$\quad$ Merge Sort$(A, r + 1, n)$
$\quad$ Merge(A, m, r, n)

The worst-case and average-case running times for the Merge sort algorithm are of order n lg n—i.e. $o(n$ lg $n)$, where n is the size of the sequence to be sorted (the average case and best case are also of order $o(n$ lg $n)$).

The Merge procedure Merges two sorted lists to produce a single sorted list. Merge (A, p, q, r) Merges A[p ... q] with A[q + 1 ... r] to yield the sorted List A [p ... r]. We use a temporary working array B[p ... r] with the same index range as A. The indices i and j point to the current element of each sub-array, and we move the smaller element into the next position in B (indicated by index k) and then increment either i or j. When we run out of entries in one array then we copy the rest of the other array into B. Finally, we copy the entire contents of B back to A.

Merge (A, p, q, r)
    Array B[p .. r]
    $i \leftarrow p$
    $k \leftarrow p$
    $j \leftarrow q{+}1$
    while ($i \leq q \wedge j \leq r$)   ... i.e., while both subarrays are non-empty

          if A[i] $\leq$ A[j]
             B[k] $\leftarrow$ A[i]
             $i \leftarrow i{+}1$
          else
             B[k] $\leftarrow$ A[j]
             $j \leftarrow j{+}1$
          $k \leftarrow k{+}1$
    while ($i \leq q$)   ... copy any leftover to B
      B[k] $\leftarrow$ A[i]
      $i \leftarrow i{+}1$
      $k \leftarrow k{+}1$
    while ($j \leq r$)   ... copy any leftover to B
      B[k] $\leftarrow$ A[j]
      $j \leftarrow j{+}1$
      $k \leftarrow k{+}1$
    for $i = p$ to r do   ... copy B back to A
      A[i] = B[i]

5.4 Binary Trees and Graph Theory

A *binary tree* (Fig. 5.6) is a tree in which each node has at most two child nodes (termed left and right child node). A node with children is termed a *parent node*, and the top node of the tree is termed the root node. Any node in the tree can be reached by starting from the root node, and by repeatedly taking either the left branch (left child) or right branch (right child) until the node is reached. Binary trees are often used in computing to implement efficient searching algorithms.

The *depth* of a node is the length of the path (i.e. the number of edges) from the root to the node. The depth of a tree is the length of the path from the root to the deepest node in the tree. A *balanced* binary tree is a binary tree in which the depth of the two subtrees of any node never differs by more than one.

Tree traversal is a systematic way of visiting each node in the tree exactly once, and we distinguish between *breadth-first search* algorithms in which every node at a particular level is visited before going to a lower level, and *depth-first search* algorithms where one starts at the root and explores as far as possible along each

Fig. 5.6 Sorted binary tree

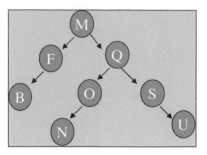

branch before backtracking. The traversal in depth-first search may be in *preorder*, *inorder* or *postorder*.

Graph algorithms are employed to solve various problems in graph theory including network cost minimization problems, construction of spanning trees, shortest path algorithms, longest path algorithms and timetable construction problems.

A length function $l : E \to \mathbb{R}$ may be defined on the edges of a connected graph $G = (V, E)$, and the shortest path from u to v in G is a path P with edge set E' such that $l(E')$ is minimal. The reader may consult the many texts on graph theory to explore many well-known graph algorithms such as Dijkstra's shortest path algorithm and longest path algorithm, Kruskal's minimal spanning tree algorithm and Prim's minimal spanning tree algorithms (Piff 1991).

5.5 Modern Cryptographic Algorithms

A cryptographic system is concerned with the secure transmission of messages. The message is encrypted prior to its transmission, and any unauthorised interception and viewing of the message is meaningless to anyone other than the intended recipient. The recipient uses a key to decrypt the encrypted text to retrieve the original message (Table 5.1).

There are essentially two different types of cryptographic systems, namely, the public key cryptosystems and secret key cryptosystems. A *public key cryptosystem* is an asymmetric cryptosystem where two different keys are employed: one for

Table 5.1 Notation in cryptography

Symbol	Description
M	Represents the message (plaintext)
C	Represents the encrypted message (cipher text)
e_k	Represents the encryption key
d_k	Represents the decryption key
E	Represents the encryption process
D	Represents the decryption process

encryption and one for decryption. The fact that a person can encrypt a message does not mean that the person is able to decrypt a message.

The same key is used for both encryption and decryption in a *secret key cryptosystem*, and anyone who has knowledge on how to encrypt messages has sufficient knowledge to decrypt messages. The encryption and decryption algorithms satisfy the following equation:

$$Dd_k(C) = Dd_k(Ee_k(M)) = M$$

There are two different keys employed in a public key cryptosystem. These are the encryption key e_k and the decryption key d_k with $e_k \neq d_k$. It is called asymmetric as the encryption key differs from the decryption key.

A symmetric key cryptosystem (Fig. 5.7) uses the same secret key for encryption and decryption, and so the sender and the receiver first need to agree on a shared key prior to communication. This needs to be done over a secure channel to ensure that the shared key remains secret. Once this has been done, they can begin to encrypt and decrypt messages using the secret key.

The encryption of a message is in effect a transformation from the space of messages m to the space of cryptosystems $\mathbb{C}$. That is, the encryption of a message with key k is an invertible transformation f such that

$$f: m \xrightarrow{k} \mathbb{C}$$

The cipher text is given by $C = E_k(M)$, where $M \in m$ and $C \in \mathbb{C}$. The legitimate receiver of the message knows the secret key k (as it will have transmitted previously over a secure channel), and so the cipher text C can be decrypted by the inverse transformation f^{-1} defined by

$$f^{-1}: \mathbb{C} \xrightarrow{k} m$$

Therefore, we have that $D_k(C) = D_k(E_k(M)) = M$ the original plaintext message.

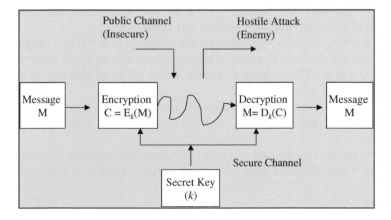

Fig. 5.7 Symmetric key cryptosystem

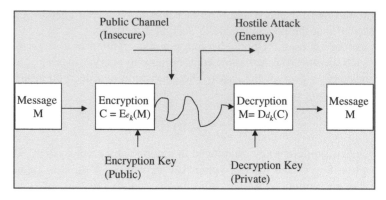

Fig. 5.8 Public key cryptosystem

A public key cryptosystem (Fig. 5.8) is an asymmetric key system where there is a separate key e_k for encryption and d_k decryption with $e_k \neq d_k$. Martin Hellman and Whitfield Diffie invented it in 1976. The fact that a person is able to encrypt a message does not mean that the person has sufficient information to decrypt messages. There is more information on cryptography in (O'Regan 2017).

5.6   Computational Complexity

An algorithm is of little practical use if it takes millions of years to compute the solution to a problem. That is, the fact that there is an algorithm to solve a problem is not sufficient, as there is also the need to consider the efficiency of the algorithm. The security of the RSA encryption algorithm relies on the fact that there is no known efficient algorithm to determine the prime factors of a large number.

There are often slow and fast algorithms for the same problem, and a measure of the complexity of an algorithm is the number of steps in its computation. An algorithm is of *time complexity* $f(n)$ if for all n and all inputs of length n, the execution of the algorithm takes at most $f(n)$ steps.

An algorithm is said to be *polynomially bounded* if there is a polynomial p (n) such that for all n and all inputs of length n the execution of the algorithm takes at most $p(n)$ steps. The notation P is used for all problems that can be solved in polynomial time.

A problem is said to be *computationally intractable* if it may not be solved in polynomial time: that is, there is no known algorithm to solve the problem in polynomial time.

A problem L is said to be in the set *NP* (non-deterministic polynomial time problems) if any given solution to L can be verified quickly in polynomial time.

A non-deterministic Turing machine may have several possibilities for its behaviour, and an input may give rise to several computations.

A problem is *NP complete* if it is in the set *NP* of non-deterministic polynomial time problems and it is also in the class of *NP hard* problems. A key characteristic to *NP* complete problems is that there is no known fast solution to them, and the time required to solve the problem using known algorithms increases quickly as the size of the problem grows. Often, the time required to solve the problem is in billions of years. That is, although any given solution may be verified quickly, there is no known efficient way to find a solution.

5.7 Review Questions

1. What is an algorithm?
2. Explain why the efficiency of an algorithm is important.
3. Investigate the principles underlying modern cryptography, and how it is related to computer algorithms.
4. What factors should be considered in the choice of algorithm where several algorithms exist for solving the particular problem?
5. Investigate some of the early algorithms developed by the Babylonians (e.g. finding square roots and factorization).
6. Explain the difference between the insertion sort algorithm and Merge sort.
7. Investigate famous computer algorithms such as Dijkstra's shortest path, Prim's algorithm and Kruskal's algorithm.

5.8 Summary

This chapter gave a short introduction to computer algorithms, where an algorithm is a well-defined procedure for solving a problem. It consists of a sequence of steps that take a set of input values and produce a set of output values. It is an exact specification of how to solve the problem, and a computer program implements the algorithm in some programming language.

It is essential that the algorithm is correct, and that it terminates in a reasonable period of time. There may be several algorithms for a problem, and so the choice of the best algorithm (e.g. fastest/most efficient) needs to be considered.

This may require mathematical analysis of the algorithm to demonstrate its correctness and efficiency, and to show that it terminates in a finite period of time. An algorithm may be implemented by a computer program, and the speed of the program depends on the algorithm employed, the input value(s), how the algorithm has been implemented in the programming language, the compiler, the operating system and the computer hardware.

An algorithm may be described in natural language, *pseudocode,* in flowchart, in a programming language or in a formal specification language.

Chapter 6
A Concise Introduction to Logic

> **Key Topics**
>
> Syllogistic Logic
> Fallacies
> Stoic Logic
> Boole and Frege
> Propositions
> Truth Tables
> Semantic Tableaux
> Natural Deduction
> Proof
> Predicates
> Universal Quantifiers
> Existential Quantifiers

6.1 Introduction

Logic is the study of reasoning and the validity of arguments, and it is concerned with the truth of statements (propositions) and the nature of truth. Formal logic is concerned with the form of arguments and the principles of valid inference. Valid arguments are truth preserving, and for a valid deductive argument, the conclusion will always be true if the premises are true.

Propositional logic is the study of propositions, where a proposition is a statement that is either true or false. Propositions may be combined with other propositions (with a logical connective) to form compound propositions. Truth tables are used to give operational definitions of the most important logical connectives, and they provide a mechanism to determine the truth-values of more complicated logical expressions.

Propositional logic may be used to encode simple arguments that are expressed in natural language, and to determine their validity. The validity of an argument may be determined from truth tables, or using inference rules such as modus ponens to establish the conclusion via deductive steps.

© Springer International Publishing AG, part of Springer Nature 2018
G. O'Regan, *World of Computing*,
https://doi.org/10.1007/978-3-319-75844-2_6

Predicate logic allows complex facts about the world to be represented, and new facts may be determined via deductive reasoning. Predicate Calculus includes predicates, variables and quantifiers, and a *predicate* is a characteristic or property that the subject of a statement can have. A predicate may include variables, and statements with variables become propositions once the variables are assigned values.

The universal quantifier is used to express a statement such as all members of the domain of discourse have property P. This is written as $(\forall x)\, P(x)$, and it expresses the statement that the property $P(x)$ is true for all x.

The existential quantifier states that there is at least one member of the domain of discourse that has property P. This is written as $(\exists x)\, P(x)$.

6.2   A Brief History of Logic

The origins of logic are with the Greeks who were interested in the nature of truth. The sophists (e.g. Protagoras and Gorgias) were teachers of rhetoric, who taught their pupils techniques in winning an argument and persuading an audience. Plato explores the nature of truth in some of his dialogues, and he is critical of the position of the sophists who argue that there is no absolute truth, and that truth instead is always relative to some frame of reference. The classic sophist position is stated by Protagoras *'Man is the measure of all things: of things which are, that they are, and of things which are not, that they are not'*. In other words, what is true for you is true for you, and what is true for me is true for me.

Socrates had a reputation for demolishing an opponent's position, and the Socratic enquiry consisted of questions and answers in which the opponent would be led to a conclusion incompatible with his original position. The approach was like a *reductio ad absurdum* argument, although Socrates was a moral philosopher who did no theoretical work on logic.

Aristotle did important work on logic, and he developed a system of logic, *syllogistic logic*, that remained in use up to the nineteenth century. Syllogistic logic is a 'term-logic', with letters used to stand for the individual terms. A syllogism consists of two premises and a conclusion, where the conclusion is a valid deduction from the two premises. Aristotle also did some early work on modal logic and was the founder of the field.

The Stoics developed an early form of propositional logic, where the assertibles (propositions) have a truth-value such that at any time they are either true or false. The assertibles may be simple or non-simple, and various connectives such as conjunctions, disjunctions and implication are used in forming more complex assertibles.

George Boole developed his symbolic logic in the mid-1800s, and it later formed the foundation for digital computing. Boole argued that logic should be considered as a separate branch of mathematics, rather than as a part of philosophy. He argued that there are mathematical laws to express the operation of reasoning in the human mind, and he showed how Aristotle's syllogistic logic could be reduced to a set of algebraic equations.

Logic plays a key role in reasoning and deduction in mathematics, but it is considered a separate discipline to mathematics. There were attempts in the early twentieth century to show that all mathematics can be derived from formal logic, and that the formal system of mathematics would be complete, with all the truths of mathematics provable in the system. However, this program failed when the Austrian logician, Kurt Goedel, showed that there are truths in the formal system of arithmetic that cannot be proved within the system (i.e. first-order arithmetic is *incomplete*).

6.2.1   Syllogistic Logic

Early work on logic was done by Aristotle in the fourth century B.C. in the *Organon* (Ackrill 1994). Aristotle regarded logic as a useful tool of enquiry into any subject, and he developed *syllogistic logic*. This is a form of reasoning where a conclusion is deduced from two premises, where each premise is in a subject–predicate form. A common or middle term is present in each of the two premises but not in the conclusion. For example,

All Greeks are mortal
Socrates is a Greek

— — — — — — — — —

Therefore, Socrates is mortal

The common (or middle) term in this example is 'Greek'. It occurs in both premises but not in the conclusion. The above argument is valid, and Aristotle studied and classified the various types of syllogistic arguments to determine those that were valid or invalid. Each premise contains a subject and a predicate, and the middle term may act as subject or a predicate. Each premise is a positive or negative affirmation, and the affirmation may be universal or particular (Table 6.1).

This leads to four basic forms of syllogistic arguments (Table 6.2) where the middle is the subject of both premises, the predicate of both premises and the subject of one premise and the predicate of the other premise.

Table 6.1 Types of syllogistic premises

Type	Symbol	Example
Universal affirmative	G A M	All Greeks are mortal
Universal negative	G E M	No Greek is mortal
Particular affirmative	G I M	Some Greeks are mortal
Particular negative	G O M	Some Greeks are not mortal

Table 6.2 Forms of syllogistic premises

	Form (i)	Form (ii)	Form (iii)	Form (iv)
Premise 1	M P	P M	P M	M P
Premise 2	M S	S M	M S	S M
Conclusion	S P	S P	S P	S P

There are four types of premises (A, E, I, O) and therefore 16 sets of premise pairs for each of the forms above. However, only some of these premise pairs will yield a valid conclusion. Aristotle went through every possible premise pair to determine if a valid argument may be derived. The syllogistic argument above is of the form (iv) and is valid:

$$
\begin{array}{c}
\text{G A M} \\
\text{S I G} \\
\hline
\text{S I M}
\end{array}
$$

Syllogistic logic is a '*term-logic*' with letters used to stand for the individual terms. Syllogistic logic was the first attempt at a science of logic and it remained in use up to the nineteenth century. There are many limitations to what it may express, and on its suitability for logical deduction.

6.2.2   Paradoxes and Fallacies

A paradox is a statement that apparently contradicts itself, and it presents a situation that appears to defy logic. Some logical paradoxes have a solution, whereas others are contradictions or invalid arguments. They often arise due to self-reference in which one or more statements refer to each other. The *liar paradox* and the *sorites paradox* were invented by Eubulides of Miletus, and the *barber paradox* was introduced by Russell to explain the contradictions in naïve set theory.

The *liar paradox* is illustrated by the statement 'Everything that I say is false', which is made by the liar. This looks like a normal sentence but it is also saying something about itself as a sentence. If the statement is true, then the statement must be false, since the meaning of the sentence is that every statement (including the current statement) made by the liar is false. If the current statement is false, then the statement that everything that I say is false is false, and so this must be a true statement.

The *Epimenides paradox* is a variant of the liar paradox. Epimenides was a Cretan who allegedly stated that '*All Cretans are liars*'. If the statement is true, then since Epimenides is Cretan, he must be a liar, and so the statement is false and we have a contradiction. However, if we assume that the statement is false and that Epimenides is lying about all Cretan being liars, then we may deduce (without contradiction) that there is at least one Cretan who is truthful. Therefore, in this case, the paradox can be avoided.

The *sorites paradox* (paradox of the heap) involves a heap of sand in which grains are individually removed. It is assumed that removing a single grain of sand does not turn a heap into a non-heap, and the paradox is to consider what happens after when the process is repeated often enough. Is a single remaining grain a heap? When does it change from being a heap to a non-heap? This paradox may be avoided by specifying a fixed boundary of the number of grains of sand required to form a heap, or to define a heap as a collection of multiple grains (≥ 2 grains). Then, any collection of grains of sand less than this boundary is not a heap.

The *barber paradox* is a variant of Russell's paradox (a contradiction in naïve set theory), which was discussed in Chap. 4. In a village, there is a barber who shaves everyone who does not shave himself, and no one else. Who shaves the barber? The answer to this question results in a contradiction, as the barber cannot shave himself, since he shaves only those who do not shave themselves. Further, as the barber does not shave himself then he falls into the group of people who would be shaved by the barber (himself). Therefore, we conclude that there is no such barber.

The purpose of a debate is to convince an audience of the correctness of your position, and to challenge and undermine your opponent's argument. Often, the arguments made are factual, but occasionally individuals skilled in rhetoric introduce bad arguments to persuade the audience. Aristotle studied and classified bad arguments (known as *fallacies*), and these include the *ad hominem* argument, the *appeal to authority* argument and the *straw man* argument (Table 6.3).

6.2.3   Stoic Logic

The Stoic school[1] was founded in the Hellenistic period by Zeno of Citium (in Cyprus) in the late fourth/early third century B.C. (Fig. 6.1). The school presented its philosophy as a way of life, and it emphasised ethics as the focus of human knowledge. The Stoics stressed the importance of living a good life in harmony with nature.

The Stoics recognised the importance of reason and logic, and Chrysippus, the head of the Stoics in the third century B.C., developed an early version of propositional logic. This was a system of deduction in which the smallest unanalyzed expressions are assertibles (Stoic equivalent of propositions). The assertibles have a truth-value such that at any moment of time they are either true or false. True

---

[1]The origin of the word Stoic is from the *Stoa Poikile* (Στοα Ποικιλη), which was a covered walkway in the Agora of Athens. Zeno taught his philosophy in a public space at this location, and his followers became known as Stoics.

Table 6.3   Table: Fallacies in arguments

Fallacy	Description/example
Hasty/accident generalisation	This is a bad argument that involves a generalisation that disregards exceptions
Slippery slope	This argument outlines a chain reaction leading to a highly undesirable situation that will occur if a certain situation is allowed. The claim is that even if one step is taken onto the slippery slope then we will fall all the way down to the bottom
Against the person (ad hominem)	The focus of this argument is to attack the person rather than the argument that the person has made
Appeal to people (*ad populum*)	This argument involves an appeal to popular belief to support an argument, with a claim that most of the population supports this argument. However, popular opinion is not always correct
Appeal to authority (*ad verecundiam*)	This argument is when an appeal is made to an authoritative figure to support an argument, and where the authority is not an expert in this area
Appeal to pity (*ad misericordiam*)	This is where the arguer tries to get people to accept a conclusion by making them feel sorry for someone
Appeal to ignorance	The arguer makes the case that there is no conclusive evidence on the issue at hand and that therefore, his conclusion should be accepted
Straw man argument	The arguer sets up a version of an opponent's position of the argument and defeats this watered-down version
Begging the question	This is a circular argument where the arguer relies on a premise that says the same thing as the conclusion and without providing any real evidence for the conclusion
Red Herring	The arguer goes off on a tangent that has nothing to do with the argument in question
False dichotomy	The arguer presents the case that there are only two possible outcomes (often there are more). One of the possible outcomes is then eliminated leading to the desired outcome

assertibles are viewed as facts in the Stoic system of logic, and false assertibles are defined as the contradictories of true ones.

Truth is temporal and assertions may change their truth-value over time. The assertibles may be simple or non-simple (more than one assertible), and there may be present tense, past tense and future tense assertibles. Chrysippus distinguished between simple and compound propositions, and he introduced a set of logical connectives for conjunction, disjunction and implication that are used to form non-simple assertibles from existing assertibles.

The conjunction connective is of the form '*both ... and ...*', and it has two conjuncts. The disjunction connective is of the form '*either ... or ... or ...*', and it consists of two or more disjuncts. Conditionals are formed from the connective '*if ...,....*' and they consist of an antecedent and a consequence.

His deductive system included various logical argument forms such as *modus ponens* and *modus tollens*. His propositional logic differed from syllogistic logic, in

Fig. 6.1 Zeno of Citium

that the Stoic logic was based on propositions (or statements) as distinct from Aristotle's term-logic. However, he could express the universal affirmation in syllogistic logic (e.g. All As are B) by rephrasing it as a conditional statement that if something is A then it is B.

Chrysippus' propositional logic did not replace Aristotle's syllogistic logic, and syllogistic logic remained in use up to the mid-nineteenth century, until George Boole developed his symbolic logic in the mid-1800s.

6.2.4   Boole's Symbolic Logic

George Boole (Fig. 2.6) was born in Lincoln, England in 1815, and he was self-taught in mathematics and Greek. He taught at various schools near Lincoln, and he developed his mathematical knowledge by working his way through Newton's Principia, as well as applying himself to the work of mathematicians such as Laplace and Lagrange.

He published regular papers from his early twenties on probability theory, differential equations and finite differences. He developed his symbolic algebra, which is the foundation for modern computing, and he is considered (along with Babbage) to be one of the grandfathers of computing. His work was theoretical, and he never actually built a computer or calculating machine. However, his symbolic logic was the perfect mathematical model for switching theory and for the design of digital circuits.

He published 'Mathematical Analysis of Logic' in 1847 (Boole 1848), and this short book developed novel ideas on a logical method. He argued that logic should be considered a separate branch of mathematics, and he showed that there are mathematical laws to express the operation of reasoning in the human mind. He showed how Aristotle's syllogistic logic could be reduced to a set of algebraic equations.

He introduced two quantities '0' and '1', with the quantity 1 used to represent the universe of thinkable objects (i.e. the universal set), and the quantity 0 represents the absence of any objects (i.e. the empty set). He then employed symbols such as x, y, z, etc., to represent collections or classes of objects given by the meaning attached to adjectives and nouns. Next, he introduced three operators (+, − and ×) that combined classes of objects.

Boole's logic appeared to have no practical use, but this changed with Claude Shannon's 1937 Master's thesis, which showed its applicability to switching theory and to the design of digital circuits. Boole's logic was discussed in Chap. 2.

6.2.5   Frege

Gottlob Frege (Fig. 6.2) was a German mathematician and logician who is considered (along with Boole) to be one of the founders of modern logic. He also made important contributions to the foundations of mathematics, and he attempted to show that all the basic truths of mathematics (or at least of arithmetic) could be derived from a limited set of logical axioms (this approach is known as *logicism*).

He invented predicate logic and the universal and existential quantifiers, and predicate logic was a significant advance on Aristotle's syllogistic logic. Predicate logic is described in more detail later in the chapter.

Frege published several important books on logic, including *Begriffsschrift* (Definition) in 1879, *Die Grundlagen der Arithmetik* (The Foundations of Arithmetic) in 1884 and the two-volume work *Grundgesetze der Arithmetik* (Basic Laws of Arithmetic), which were published in 1893 and 1903. These books described his invention of axiomatic predicate logic, the use of quantified variables and the application of his logic to the foundations of arithmetic.

He used his predicate logic to define the natural numbers and their properties. He had intended producing three volumes of the Basic Laws of Arithmetic, with the later volumes dealing with the real numbers and their properties. However, Bertrand Russell discovered a contradiction in Frege's system (see Russell's paradox in Chap. 4), which he communicated to Frege shortly before the publication of the second volume. Frege struggled to find a satisfactory solution, and Russell later introduced the theory of types in the *Principia Mathematica* as a solution.

Fig. 6.2 Gottlob Frege

6.3 Propositional Logic

Propositional logic is the study of propositions where a proposition is a statement that is either true or false. There are many examples of propositions such as '1 + 1 = 2' which is a true proposition, and the statement that 'Today is Wednesday' which is true if today is Wednesday and false otherwise. The statement $x > 0$ is not a proposition as it contains a variable x, and it is only meaningful to consider its truth or falsity only when a value is assigned to x. Once the variable x is assigned a value it becomes a proposition. The statement 'This sentence is false' is not a proposition as it contains a self-reference that contradicts itself. Clearly, if the statement is true then it is false, and if is false then it is true.

A propositional variable may be used to stand for a proposition (e.g. let the variable P stand for the proposition '2 + 2 = 4' which is a true proposition). A propositional variable takes the value true or false. The negation of a proposition P (denoted $\neg P$) is the proposition that is true if and only if P is false, and is false if and only if P is true.

A well-formed formula (*wff*) in propositional logic is a syntactically correct formula created according to the syntactic rules of the underlying Calculus. It is built up from variables, constants, terms and logical connectives such as conjunction (and), disjunction (or), implication (if … then …), equivalence (if and only if) and negation. A distinguished subset of these well-formed formulae are the *axioms* of the Calculus, and there are *rules of inference* that allow the truth of new formulae to be

Table 6.4 Truth table for formula W

A	B	$W\,(A,\,B)$
T	T	T
T	F	F
F	T	F
F	F	T

derived (from the axioms and from formulae that have already been demonstrated to be true in the Calculus).

A formula in propositional Calculus may contain several propositional variables, and the truth or falsity of the individual variables needs to be known prior to determining the truth or falsity of the logical formula.

Each propositional variable has two possible values, and a formula with n-propositional variables has 2^n values associated with the n-propositional variables. The set of values associated with the n variables may be used derive a truth table with 2^n rows and $n + 1$ columns. Each row gives each of the 2^n truth-values that the n variables may take, and column $n + 1$ gives the result of the logical expression for that set of values of the propositional variables. For example, the propositional formula W defined in the truth table (Table 6.4) has two propositional variables A and B, with $2^2 = 4$ rows for each of the values that the two propositional variables may take. There are $2 + 1 = 3$ columns with W defined in the third column.

A rich set of connectives is employed in the Calculus to combine propositions and to build up the well-formed formulae. This includes the conjunction of two propositions $(A \land B)$, the disjunction of two propositions $(A \lor B)$ and the implication of two propositions $(A \rightarrow B)$. These connectives allow compound propositions to be formed, and the truth of the compound propositions is determined from the truth-values of its constituent propositions and the rules associated with the logical connective. The meaning of the logical connectives is given by truth tables.[2]

Mathematical logic is concerned with inference, and it involves proceeding in a methodical way from the axioms and using the rules of inference to derive further truths.

The rules of inference allow new propositions to be deduced from a set of existing propositions. A valid argument (or deduction) is truth preserving: i.e. for a valid logical argument if the set of premises is true then the conclusion (i.e. the deduced proposition) must be true. The rules of inference include rules such as *modus ponens*, and this rule states that given the truths of the proposition A, and the proposition $A \rightarrow B$, then the truth of proposition B may be deduced.

The propositional Calculus is employed in reasoning about propositions, and it may be applied to formalise arguments in natural language. *Boolean algebra* is used in computer science, and it is named after George Boole, who was the first professor of mathematics at Queens College, Cork.[3]

---

[2]Basic truth tables were first used by Frege and developed further by Post and Wittgenstein.

[3]This institution is now known as University College Cork.

6.3.1   Truth Tables

Truth tables give operational definitions of the most important logical connectives, and they provide a mechanism to determine the truth-values of more complicated compound expressions. Compound expressions are formed from propositions and connectives, and the truth-values of a compound expression containing several propositional variables are determined from the underlying propositional variables and the logical connectives.

The conjunction of A and B (denoted $A \wedge B$) is true if and only if both A and B are true, and is false in all other cases (Table 6.5). The disjunction of two propositions A and B (denoted $A \vee B$) is true if at least one of A and B are true, and false in all other cases (Table 6.6). The disjunction operator is known as the '*inclusive or*' operator as it is also true when both A and B are true; there is also an *exclusive or* operator that is true exactly when one of A or B is true and is false otherwise.

Example 6.1 Consider proposition A given by 'An orange is a fruit' and the proposition B given by '$2 + 2 = 5$' then A is true and B is false. Therefore,

 (i) $A \wedge B$ (i.e. an orange is a fruit and $2 + 2 = 5$) is false.
(ii) $A \vee B$ (i.e. an orange is a fruit or $2 + 2 = 5$) is true.

The implication operation ($A \rightarrow B$) is true if whenever A is true means that B is also true, and it is also true whenever A is false (Table 6.7). It is equivalent (as shown by a truth table) to $\neg A \vee B$. The equivalence operation ($A \leftrightarrow B$) is true whenever both A and B are true, or whenever both A and B are false (Table 6.8).

The not operator ($\neg$) is a unary operator (i.e. it has one argument) and is such that $\neg A$ is true when A is false and is false when A is true (Table 6.9).

Example 6.2 Consider proposition A given by 'Jaffa cakes are biscuits' and the proposition B given by '$2 + 2 = 5$', then A is true and B is false. Therefore,

  (i) $A \rightarrow B$ (i.e. Jaffa cakes are biscuits implies $2 + 2 = 5$) is false.
 (ii) $A \leftrightarrow B$ (i.e. Jaffa cakes are biscuits is equivalent to $2 + 2 = 5$) is false.
(iii) $\neg B$ (i.e. $2 + 2 \neq 5$) is true.

Creating a Truth Table
The truth table for a well-formed formula $W(P_1, P_2, \ldots, P_n)$ is a table with 2^n rows and $n + 1$ columns. Each row lists a different combination of truth-values of the propositions $P_1, P_2, \ldots, P_n$ followed by the corresponding truth-value of W.

Table 6.10 gives the truth table for a formula W with three propositional variables (meaning that there are $2^3 = 8$ rows in the truth table and $3 + 1 = 4$ columns).

Table 6.5 Conjunction

A	B	$A \wedge B$
T	T	T
T	F	F
F	T	F
F	F	F

Table 6.6 Disjunction

A	B	$A \vee B$
T	T	T
T	F	T
F	T	T
F	F	F

Table 6.7 Implication

A	B	$A \rightarrow B$
T	T	T
T	F	F
F	T	T
F	F	T

Table 6.8 Equivalence

A	B	$A \leftrightarrow B$
T	T	T
T	F	F
F	T	F
F	F	T

Table 6.9 Not operation

A	$\neg A$
T	F
F	T

Table 6.10 Truth table for W (P, Q, R)

P	Q	R	$W(P, Q, R)$
T	T	T	F
T	T	F	F
T	F	T	F
T	F	F	T
F	T	T	T
F	T	F	F
F	F	T	F
F	F	F	F

6.3.2 Properties of Propositional Calculus

There are many well-known properties of the propositional Calculus such as the commutative, associative and distributive properties. These ease the evaluation of complex expressions and allow logical expressions to be simplified.

The *commutative property* holds for the conjunction and disjunction operators, and it states that the order of evaluation of the two propositions may be reversed without affecting the resulting truth-value. The *associative property* holds for the conjunction and disjunction operators, which means that order of evaluation of a sub-expression does not affect the resulting truth-value. The conjunction operator *distributes* over the disjunction operator and vice versa.

The result of the logical conjunction of two propositions is false if one of the propositions is false (irrespective of the value of the other proposition). The result of the logical disjunction of two propositions is true if one of the propositions is true (irrespective of the value of the other proposition). The result of the logical disjunction of two propositions, where one of the propositions is known to be false is given by the truth-value of the other proposition. The result of the logical conjunction of two propositions, where one of the propositions is known to be true, is given by the truth-value of the other proposition.

The conjunction and disjunction operators are *idempotent*. That is, when the arguments of the conjunction or disjunction operator are the same proposition A the result is A. The *law of the excluded middle* is a fundamental property of the propositional Calculus. It states that a proposition A is either true or false: i.e. there is no third logical value.

We noted that $A \rightarrow B$ is logically equivalent to $\neg A \vee B$ (same truth table), and clearly $\neg A \vee B$ is the same as $\neg A \vee \neg\neg B$ which is the same as $\neg\neg B \vee \neg A$ which is logically equivalent to $\neg B \rightarrow \neg A$. That is, $A \rightarrow B$ is logically equivalent to $\neg B \rightarrow \neg A$ (this is known as the *contrapositive*). De Morgan's law states:

$$\neg(A \wedge B) \equiv \neg A \vee \neg B$$
$$\neg(A \vee B) \equiv \neg A \wedge \neg B$$

A proposition that is true for all values of its constituent propositional variables is termed a *tautology*. For example, the proposition $A \vee \neg A$ is a tautology. A proposition that is false for all values of its constituent propositional variables is termed a *contradiction*. For example, the proposition $A \wedge \neg A$ is a contradiction.

6.3.3 Proof in Propositional Calculus

Logic enables further truths to be derived from existing truths by rules of inference that are truth preserving. Propositional Calculus is both *complete* and *consistent*. The completeness property means that all true propositions are deducible in the

Calculus, and the consistency property means that there is no formula A such that both A and $\neg A$ are deducible in the Calculus.

An argument in propositional logic consists of a sequence of formulae that are the premises of the argument and a further formula that is the conclusion of the argument. One elementary way to see if the argument is valid is to produce a truth table to determine if the conclusion is true whenever all the premises are true.

Consider a set of premises P_1, P_2, ... P_n and conclusion Q. Then to determine if the argument is valid using a truth table involves adding a column in the truth table for each premise P_1, P_2, ... P_n, and then identifying the rows in the truth table for which these premises are all true. The truth-value of the conclusion Q is examined in each of these rows, and if Q is true for each case for which P_1, P_2, ... P_n are all true then the argument is valid. This is equivalent to $P_1 \wedge P_2 \wedge ... \wedge P_n \rightarrow Q$ is a tautology.

An alternative approach is to assume the negation of the desired conclusion (i.e. $\neg Q$) and to show that the premises and the negation of the conclusion result in a contradiction (i.e. $P_1 \wedge P_2 \wedge ... \wedge P_n \wedge \neg Q$) is a contradiction. The use of truth tables becomes cumbersome when there are many variables involved, as there are 2^n truth table entries for n propositional variables.

Procedure for Proof by Truth Table

(i)   Consider argument P_1, P_2, ..., P_n with conclusion Q.
(ii)   Draw truth table with column in truth table for each premise P_1, P_2, ..., P_n.
(iii) Identify the rows in truth table for when the premises are all true.
(iv)   Examine truth-value of Q for these rows.
(v)   If Q is true for each case that P_1, P_2, ... P_n are true then the argument is valid.
(vi)   That is $P_1 \wedge P_2 \wedge ... \wedge P_n \rightarrow Q$ is a tautology.

Example 6.3 (Truth Tables) Consider the argument adapted from (Kelly 1997) and determine if it is valid.

If the pianist plays the concerto then crowds will come if the prices are not too high.

If the pianist plays the concerto then the prices will not be too high.

Therefore, if the pianist plays the concerto then crowds will come.

Solution
We will adopt a common proof technique that involves showing that the negation of the conclusion is incompatible (inconsistent) with the premises, and from this, we deduce the conclusion must be true.

Let P stand for 'The pianist plays the concerto'; C stands for 'Crowds will come'; and H stands for 'Prices are too high'. Then, the argument may be expressed in propositional logic as

$$P \rightarrow (\neg H \rightarrow C)$$
$$P \rightarrow \neg H$$
$$P \rightarrow C$$

Table 6.11 Proof of argument with a truth table

P	C	H	¬H	¬H→C	P→(¬H→ C)	P→¬H	P→C	¬(P→C)	*
T	T	T	F	T	T	F	T	F	F
T	T	F	T	T	T	T	T	F	F
T	F	T	F	T	T	F	F	T	F
T	F	F	T	F	F	T	F	T	F
F	T	T	F	T	T	T	T	F	F
F	T	F	T	T	T	T	T	F	F
F	F	T	F	T	T	T	T	F	F
F	F	F	T	F	T	T	T	F	F

Then, we negate the conclusion $P \rightarrow C$ and check the consistency of $P \rightarrow (\neg H \rightarrow C) \wedge (P \rightarrow \neg H) \wedge \neg (P \rightarrow C)^*$ using a truth table (Table 6.11).

It is clear from the last column in the truth table that the negation of the conclusion is incompatible with the premises, and therefore, it cannot be the case that the premises are true and the conclusion false. Therefore, the conclusion must be true whenever the premises are true, and we conclude that the argument is valid.

6.3.4 Semantic Tableaux in Propositional Logic

The problem with using truth tables to prove the validity of an argument is that they can get extremely large very quickly, and so we will consider an alternative approach known as semantic tableaux.

The basic idea of semantic tableaux (developed by the Dutch logician, Evert Beth) is to determine if it is possible for a conclusion to be false when all the premises are true. If this is not possible, then the conclusion must be true when the premises are true, and so the conclusion is *semantically entailed* by the premises. The method of semantic tableaux is a technique to expose inconsistencies in a set of logical formulae, by identifying conflicting logical expressions.

We present a short summary of the rules of semantic tableaux in Table 6.12, and we then give a proof of Example 6.3 using semantic tableaux.

Whenever a logical expression A and its negation $\neg A$ appear in a branch of the tableau, then an inconsistency has been identified in that branch, and the branch is said to be *closed*. If all the branches of the semantic tableaux are closed, then the logical propositions from which the tableau was formed are mutually inconsistent, and cannot be true together.

The method of proof is to negate the conclusion and to show that all branches in the semantic tableau are closed, and that therefore, it is not possible for the premises of the argument to be true and for the conclusion to be false. Therefore, the argument is valid and the conclusion follows from the premises.

Table 6.12 Rules of semantic tableaux

Rule No.	Definition	Description
1.	$A \wedge B$ A B	If $A \wedge B$ is true then both A and B are true, and may be added to the branch containing $A \wedge B$
2.	$A \vee B$ $A \qquad B$	If $A \vee B$ is true then either A or B is true, and we add two new branches to the tableaux, one containing A and one containing B
3.	$A \rightarrow B$ $\neg A \qquad B$	If $A \rightarrow B$ is true then either $\neg A$ or B is true, and we add two new branches to the tableaux, one containing $\neg A$ and one containing B
4.	$A \leftrightarrow B$ $A \wedge B \qquad \neg A \wedge \neg B$	If $A \leftrightarrow B$ is true then either $A \wedge B$ or $\neg A \wedge \neg B$ is true, and we add two new branches, one containing $A \wedge B$ and one containing $\neg A \wedge \neg B$
5.	$\neg\neg A$ A	If $\neg\neg A$ is true then A may be added to the branch containing $\neg\neg A$
6.	$\neg(A \wedge B)$ $\neg A \qquad \neg B$	If $\neg(A \wedge B)$ is true then either $\neg A$ or $\neg B$ is true, and we add two new branches to the tableaux, one containing $\neg A$ and one containing $\neg B$
7.	$\neg(A \vee B)$ $\neg A$ $\neg B$	If $\neg(A \vee B)$ is true then both $\neg A$ and $\neg B$ are true, and may be added to the branch containing $\neg(A \vee B)$
8.	$\neg(A \rightarrow B)$ A $\neg B$	If $\neg(A \rightarrow B)$ is true then both A and $\neg B$ are true, and may be added to the branch containing $\neg(A \rightarrow B)$

Example 6.4 (Semantic Tableaux) Perform the proof, for example, 6.3 using semantic tableaux.

Solution

We formalised the argument previously as

(Premise 1)	$P \rightarrow (\neg H \rightarrow C)$
(Premise 2)	$P \rightarrow \neg H$
(Conclusion)	$P \rightarrow C$

We negate the conclusion to get $\neg(P \rightarrow C)$ and we show that all branches in the semantic tableau are closed, and that therefore, it is not possible for the premises of the argument to be true and for the conclusion false.

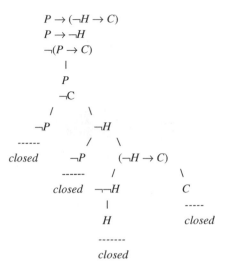

We have showed that all branches in the semantic tableau are closed, and that therefore, it is not possible for the premises of the argument to be true and for the conclusion false. Therefore, the argument is valid as required.

6.3.5   Natural Deduction

Gerhard Gentzen (Fig. 6.3) developed a method for logical deduction known as 'Natural Deduction', which aims to be as close as possible to natural reasoning. Gentzen worked as an assistant to David Hilbert at the University of Göttingen in Germany, and he died in Prague at the end of the Second World War.

Natural deduction includes rules for $\land$, $\lor$, $\rightarrow$ introduction and elimination and also for *reductio ad absurdum*. There are ten inference rules in the system, with two inference rules per operator (an introduction rule and an elimination rule). Natural deduction is described in more detail in O'Regan (2017).

6.3.6   Applications of Propositional Calculus

Propositional Calculus may be employed in reasoning with arguments in natural language. First, the premises and conclusion of the argument are identified and formalised into propositions, and we then determine if the conclusion is a valid deduction from the premises. Consider the following argument that aims to prove that Superman does not exist.

'If Superman were able and willing to prevent evil, he would do so. If Superman were unable to prevent evil he would be impotent; if he were unwilling to prevent

Fig. 6.3 Gerhard Gentzen

evil he would be malevolent; Superman does not prevent evil. If superman exists he is neither malevolent nor impotent; therefore, Superman does not exist'.

First, letters are employed to represent the propositions as follows:

a   Superman is able to prevent evil.
w   Superman is willing to prevent evil.
i   Superman is impotent.
m   Superman is malevolent.
p   Superman prevents evil.
e   Superman exists.

Then, the argument above is formalised in propositional logic as follows:

$$
\begin{array}{ll}
\textit{Premises} & \\
P_1 & (a \wedge w) \to p \\
P_2 & (\neg a \to i) \wedge (\neg w \to m) \\
P_3 & \neg p \\
P_4 & e \to i \wedge \neg m \\
\hline
\textit{Conclusion} & \neg e
\end{array}
$$

Proof that Superman Does Not Exist

1.	$\neg p$	P_3
2.	$\neg(a \wedge w) \vee p$	P_1 ($A \rightarrow B \equiv \neg A \vee B$)
3.	$\neg(a \wedge w)$	1, 2 $A \vee B, \neg B \vdash A$
4.	$\neg a \vee \neg w$	3, De Morgan's law
5.	$(\neg a \rightarrow i)$	P_2 ($\wedge$-Elimination)
6.	$\neg a \rightarrow i \vee m$	5, $x \rightarrow y \vdash x \rightarrow y \vee z$
7.	$(\neg w \rightarrow m)$	P_2 ($\wedge$-Elimination)
8.	$\neg w \rightarrow i \vee m$	7, $x \rightarrow y \vdash x \rightarrow y \vee z$
9.	$(\neg a \vee \neg w) \rightarrow (i \vee m)$	8, $x \rightarrow z, y \rightarrow z \vdash x \vee y \rightarrow z$
10.	$(i \vee m)$	4,9 Modus Ponens
11.	$e \rightarrow \neg(i \vee m)$	P_4 (De Morgan's Law)
12.	$\neg e \vee \neg (i \vee m)$	11, ($A \rightarrow B \equiv \neg A \vee B$)
13.	$\neg e$	10, 12 $A \vee B, \neg B \vdash A$

Therefore, the conclusion that Superman does not exist is a valid deduction from the given premises. Next, we discuss predicate Calculus.

6.4 Predicate Logic

Predicate logic is a richer system than propositional logic, and it allows complex facts about the world to be represented. It allows new facts about the world to be derived in a way that guarantees that if the initial facts are true then the conclusions are true. Predicate Calculus includes predicates, variables, constants and quantifiers.

A *predicate* is a characteristic or property that an object can have, and we are predicating some property of the object. For example, '*Socrates is a Greek*' expresses the property that the subject 'Socrates' has the property that he is 'Greek'. This may be expressed as $G(s)$, with capital letters standing for predicates and small letters standing for objects. A predicate may include variables, and a statement with a variable becomes a proposition once the variables are assigned values. For example, $G(x)$ states that the variable x is a Greek, whereas $G(s)$ is an assignment of values to x. The set of values that the variables may take is termed the universe of discourse, and the variables take values from this set.

Predicate Calculus employs quantifiers to express properties such as all members of the domain have a particular property: e.g. $(\forall x)P(x)$, or that there is at least one member that has a particular property: e.g. $(\exists x)P(x)$. These are referred to as the *universal and existential quantifiers*.

The syllogism 'All Greeks are mortal; Socrates is a Greek; therefore, Socrates is mortal' may be easily expressed in predicate Calculus by

$$(\forall x)(G(x) \rightarrow M(x))$$
$$G(s)$$
$$-\ -\ -\ -\ -\ -\ -\ -\ -\ -$$
$$M(s)$$

In this example, the predicate $G(x)$ stands for x is a Greek and the predicate M (x) stands for x is mortal. The formula $G(x) \rightarrow M(x)$ states that if x is a Greek then x is mortal, and the formula $(\forall x)(G(x) \rightarrow M(x))$ states for any x that if x is a Greek then x is mortal. The formula $G(s)$ states that Socrates is a Greek and the formula M (s) states that Socrates is mortal.

Example 6.5 (Predicates) A predicate may have one or more variables. A predicate that has only one variable (i.e. a unary or 1-place predicate) is often related to sets, a predicate with two variables (a 2-place predicate) is a relation and a predicate with n variables (an n-place predicate) is an n-ary relation. Propositions do not contain variables and so they are 0-place predicates. The following are examples of predicates:

  (i)   The predicate *Prime(x)* states that x is a prime number (with the natural numbers being the universe of discourse).
  (ii) *Lawyer(a)* may stand for a is a lawyer.
  (iii) Mean(m, x, y) states that m is the mean of x and y: i.e. $m = 1/2(x + y)$
  (iv) LT(x, 6) states that x is less than 6.
  (v)   GT(x, y) states that x is greater than y.
  (vi) LE(x, y) states that x is less than or equal to y.
  (vii) Real(x) states that x is a real number.
  (viii) Father(x, y) states that x is the father of y.
  (ix) $\neg(\exists x)(Prime(x) \land B(x, 32, 36))$ states that there is no prime between 32 and 36.

Universal and Existential Quantification
The universal quantifier is used to express a statement such as that all members of the domain have property P. This is written as $(\forall x)P(x)$ and expresses the statement that the property $P(x)$ is true for all x. Similarly, $(\forall x_1, x_2, \ldots, x_n) P(x_1, x_2, \ldots, x_n)$ states that property $P(x_1, x_2, \ldots, x_n)$ is true for all $x_1, x_2, \ldots, x_n$. Clearly, the predicate $(\forall x) P(a, b)$ is identical to $P(a, b)$ since it contains no variables, and the predicate $(\forall y \in \mathbb{N}) (x \leq y)$ is true if $x = 1$ and false otherwise.

The existential quantifier states that there is at least one member in the domain of discourse that has property P. This is written as $(\exists x)P(x)$ and the predicate $(\exists x_1, x_2, \ldots, x_n) P(x_1, x_2, \ldots, x_n)$ states that there is at least one value of $(x_1, x_2, \ldots, x_n)$ such that $P(x_1, x_2, \ldots, x_n)$ is true.

Example 6.6 (Quantifiers)

(i) ($\exists p$) (Prime(p) $\land$ $p > 1,000,000$) is true
    It expresses the fact that there is at least one prime number greater than a
    million, which is true as there are an infinite number of primes.

(ii) ($\forall x$) ($\exists y$) $x < y$ is true
    This predicate expresses the fact that given any number x we can always find a
    larger number: e.g. take $y = x + 1$.

(iii) ($\exists y$) ($\forall x$) $x < y$ is false
    This predicate expresses the statement that there is a natural number y such
    that all natural numbers are less than y. Clearly, this statement is false since
    there is no largest natural number, and so the predicate is false.

Comment 6.1
*It is important to be careful with the order in which quantifiers are written, as the
meaning of a statement may be completely changed by the transposition of two
quantifiers.*

Scope of Quantifiers
The scope of the quantifier ($\forall x$) in the well-formed formula ($\forall x$) A is A. Similarly,
the scope of the quantifier ($\exists x$) in the well-formed formula ($\exists x$) B is B. The variable
x that occurs within the scope of the quantifier is said to be a *bound variable*, and a
variable that is not within the scope of a quantifier is said to be *free*.

6.4.1   Semantic Tableaux in Predicate Calculus

We discussed the use of semantic tableaux for determining the validity of arguments in propositional logic earlier in the chapter, and its approach is to negate the
conclusion of an argument and to show that this results in inconsistency with the
premises of the argument.

The use of semantic tableaux is similar with predicate logic, except that there are
some additional rules to consider. As before, if all branches of a semantic tableau
are closed, then the premises and the negation of the conclusion are mutually
inconsistent, and we deduce that the conclusion must be true. The additional rules
logic is detailed in Table 6.13.

Example 6.7 (Semantic Tableaux) Show that the syllogism 'All Greeks are mortal;
Socrates is a Greek; therefore, Socrates is mortal' is a valid argument in predicate
Calculus.

Table 6.13 Extra rules of semantic tableaux (for predicate Calculus)

Rule No.	Definition	Description
1	$(\forall x)\ A(x)$ $A(t)$ where t is a term	Universal instantiation
2.	$(\exists x)\ A(x)$ $A(t)$ where t is a term that has not been used in the derivation so far	Rule of existential instantiation. The term 't' is often a constant 'a'
3.	$\neg(\forall x)\ A(x)$ $(\exists x)\ \neg A(x)$	
4.	$\neg(\exists x)\ A(x)$ $(\forall x)\neg A(x)$	

Solution

We expressed this argument previously as $(\forall x)\ (G(x) \rightarrow M(x))$; $G(s)$; $M(s)$. Therefore, we negate the conclusion (i.e. $\neg M(s)$) and try to construct a closed tableau.

$$(\forall x)(G(x)\rightarrow M(x))$$
$$G(s)$$
$$\neg M(s).$$
$$G(s) \rightarrow M(s) \qquad\qquad\qquad \text{Universal Instantiation}$$
$$\wedge$$
$$\neg G(s) \qquad M(s)$$
$$\text{-----} \qquad \text{--------}$$
$$\textit{closed} \qquad \textit{closed}$$

Therefore, as the tableau is closed we deduce that the negation of the conclusion is inconsistent with the premises, and that therefore, the conclusion follows.

6.5 Review Questions

1. Draw a truth table to show that $\neg (P \rightarrow Q) \equiv P \wedge \neg Q$.
2. Translate the sentence 'Execution of program P begin with $x < 0$ will not terminate' into propositional form.
3. Explain the difference between the universal and the existential quantifier.
4. Express the following statements in the predicate Calculus:

   a. All natural numbers are greater than 10.
   b. There is at least one natural number between 5 and 10.

5. Which of the following predicates are true?

   a. $\forall i \in \{10, \ldots, 50\}.\ i^2 < 2000 \wedge i < 100$.
   b. $\exists\, i \in \mathbb{N}.\ i > 5 \wedge i^2 = 25$.

6. Use semantic tableaux to show that $(A \rightarrow A) \vee (B \wedge \neg B)$ is true.

6.6 Summary

Propositional logic is the study of propositions, where a proposition is a statement that is either true or false. A formula in propositional Calculus may contain several variables, and the truth or falsity of the individual variables, and the meanings of the logical connectives determine the truth or falsity of the logical formula.

A rich set of connectives is employed to build up the well-formed formulae of the Calculus. This includes the conjunction of two propositions $(A \wedge B)$, the disjunction of two propositions $(A \vee B)$ and the implication of two propositions $(A \rightarrow B)$. These connectives allow compound propositions to be formed, and the truth of the compound propositions is determined from the truth-values of the constituent propositions and the rules associated with the logical connectives. The meaning of the logical connectives is given by truth tables.

Propositional Calculus is complete and consistent with all true propositions deducible in the Calculus, and there is no formula A such that both A and $\neg A$ are deducible in the Calculus.

An argument in propositional logic consists of a sequence of formulae that are the premises and another formula that is the conclusion. One elementary way to see if the argument is valid is to produce a truth table to determine if the conclusion is true whenever the premises are true.

Predicates are statements involving variables and these statements become propositions once the variables are assigned values. Predicate Calculus allows expressions such as all members of the domain have a particular property to be expressed formally: e.g. $(\forall x)\ Px$, or that there is at least one member that has a particular property: e.g. $(\exists x)\ Px$.

Semantic tableaux may be used for determining the validity of arguments in propositional or predicate logic, and its approach is to negate the conclusion of an argument and to show that this results in inconsistency with the premises of the argument.

Chapter 7
Human–Computer Interaction

7.1 Introduction

Human–computer interaction (HCI) is a branch of computer science that is concerned with the design, evaluation and implementation of interactive computing systems for human use. It is focused on the interfaces between people and computers, and involves several different fields including computer science, cognitive psychology, design and communication. The human–computer interaction field has evolved over the decades to include text-based interaction systems, graphical user interfaces (GUI) and voice user interfaces (VUI) for speech recognition and speech synthesis.

The interaction between humans and machines was mainly limited to information technology professionals from the early days of computing up to the mid/late 1970s. This changed after the invention of the microprocessor in the early 1970s, which led to an explosion of interest from computer hobbyists, and the subsequent development of home computers from the mid-1970s. The introduction of the IBM personal computer in the early 1980s meant that everyone in the world was now was a potential computer user, and it led to a new market of personal applications and tools to support the user. However, it was clear that there were serious deficiencies with respect to the usability of computers in carrying out the tasks that users wished to perform.

Humans interact with computers in many ways, and so it is important to understand the interface between human and machines to facilitate an effective interaction. The early computer systems were *batch processing* (running programs in batches without human intervention) on a large expensive mainframe computer.

© Springer International Publishing AG, part of Springer Nature 2018

G. O'Regan, *World of Computing*,

https://doi.org/10.1007/978-3-319-75844-2_7

The interaction between the human (operator) and computer was limited, and it consisted of placing the punched cards (encoded instructions to the computer) on the card reader, and the computer would then process the cards overnight. These computers were slow and expensive, and it was important that they be used efficiently 24 h a day. The computer could run only one program at a time, and programmers were unable to interact with the computer while it was running, and this made it difficult and time-consuming to identify and correct errors.

A *text-based interface* (also known as a command line interface) is where the system interaction (input and output) and navigation are text-based. They are easier to use than punched card programming, but require skilled operators due to the difficulty in remembering long lists of system commands.

Licklider wrote an influential paper '*Man-Computer Symbiosis*' in 1960 (Licklider 1960), in which he outlined the need for a simple interaction between users and computers. This paper mentioned ideas such as sharing computers among many users, interactive information processing and programming, large-scale storage and retrieval, and speech and handwriting recognition.

Doug Engelbart was one of the main developers of NLS (oN Line System) in the late 1960s, and this online word processor system had features such as the first computer mouse, time-sharing and a command line interface. User trials and testing was employed in its development as part of a *philosophy towards a system adapting to people rather than people adapting to a system*.

One of the most well-known text-based operating systems was Microsoft's MS/DOS operating system for IBM compatible personal computers, which was introduced in 1981 (Fig. 7.1). Text-based interfaces are effective for expert users but are more difficult for users with an average level of knowledge, as they have a steep learning curve and the difficulty in remembering a long list of system commands. The fact that they are not very intuitive or user-friendly motivated research into alternative approaches.

The *graphical user interface* (GUI) is a human–computer interface that uses graphical icons, menus and windows to represent information and action to the user.

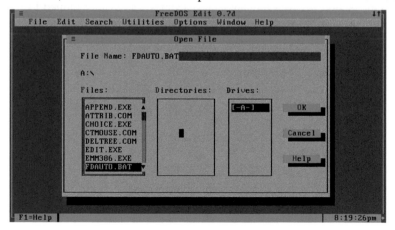

Fig. 7.1 FreeDOS text editing

It was a revolution in human and computer interaction and the GUI was intuitive and user-friendly. They have made computers and electronic devices attractive to non-technical users, and the usability of the GUI has allowed a large range of users with varying ability and expertise to successfully interact with computers.

Early work on graphical user interfaces took place at Xerox PARC in the 1970s with their work on the Xerox Alto personal workstation (Fig. 3.24). This was the first computer to use a mouse-driven graphical user interface, and it was introduced in the mid-1970s. It was essentially a small minicomputer rather than a personal computer (it was not based on the microprocessor). Its significance is that it had a major impact on the user interface design, and especially on the design of the Apple Macintosh computer.

The Xerox Star was introduced in the early 1980s, and it followed sound usability principles (prototyping and analysis, iterative development and testing with users) in its development. Steve Jobs visited Xerox PARC in late 1979, and he realised that the future of personal computing was with computers that employed a graphical user interface (such as in the Xerox Alto). Jobs was amazed that Xerox had not commercialised the technology, as he saw its graphical user interface as a revolution in computing and a potential goldmine in the future of computing. The design of the Apple Macintosh was heavily influenced by the design of the Xerox Alto, and the release of the Macintosh was a major milestone in computing.

The Macintosh was a much easier machine to use than the existing IBM personal computer. Its friendly and intuitive graphical user interface was a revolutionary change from the command-driven operating system of the IBM PC, which required the users to be familiar with its operating system commands. It was 1990 before Microsoft introduced its Windows 3.0 GUI-driven operating system (Fig. 7.2).

Today, the prevalent paradigm in human–computer interaction is the WIMP (windows, icons, menus and pointers) paradigm, which is comprised of a graphic and text interface navigated by a mouse and keyboard. The future of HCI is predicted to be the SILK (speech, image, language and knowledge) paradigm, where communication between humans and machine will be more natural and intuitive.

7.2 HCI Principles

The success of computer systems is critically influenced by the design of the human–computer interaction, and in the achievement of end-user computing satisfaction. Human–computer interaction is concerned with the study of humans and machines, and so it needs knowledge of both to be effective. The study of machines requires knowledge of computer graphics, programming languages, capabilities of current technology and so on, whereas on the human side it requires knowledge of cognitive psychology, ergonomics and other human factors such as usability and computer user satisfaction.

There are several fundamental principles and models underlying HCI. It is essential to understand the user and their characteristics, as well as their diversity in

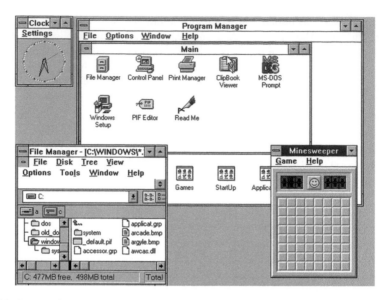

Fig. 7.2 Microsoft Windows 3.11 (1993). Used with permission from Microsoft

age, experience, physical and intellectual abilities, and so on. It is customary to distinguish between two types of user knowledge (IT and domain knowledge), and the user's proficiency in each type of knowledge yields several user categories that range between novice and expert.

– Interface knowledge (knowledge of the IT technology).
– Domain/task knowledge of the real-world system.

The software will generally support multiple user categories, where novices get opportunities to learn about the system and have fewer opportunities for error. It is important to understand the domain in which the software will be used and to identify the tasks to be performed as well as the frequency in which they will be performed.

There have been several rules and principles proposed for HCI design including Shneiderman's 'Eight Golden Rules of Interface Design' (Table 7.1) (Shneiderman and Plaisant 2005).

7.3 Software Usability and User-Centred Design

Usability has become important in software engineering and especially with the emergence of the World Wide Web in the early 1990s. The usability of the software is the perception that a user or group of users has of its quality and ease of use (i.e. is the software easy to use and easy to learn?), and its efficiency and effectiveness.

Table 7.1 Eight golden rules of interface design

Principle	Description
Strive for consistency	Consistent terminology, sequences of action and commands throughout the system
Enable frequent users to use shortcuts	The user will naturally desire to reduce the number of interactions as the frequency of use increases
Provide informative feedback	There should be appropriate system feedback
Design dialogue to yield closure	Sequences of actions should be organised into groups with a beginning, middle and end
Offer simple error handling	Design the system (as far as possible) to prevent the user from making a serious error. The system should be able to detect an error and provide a handling mechanism
Permit easy reversal of actions	This is important to the user as it means that errors can be undone
Support internal locus of control	The system should be designed to make the users initiators of actions rather than responders to actions
Reduce short-term memory load	There are limitations to human processing in short-term memory, and so displays should be kept simple

Usability is a multidisciplinary field, and psychological testing may be employed to evaluate the perception that users have of the computer system. Usability is defined in the ISO 9241 standard as:

> **Usability** is the degree to which software can be used by specified consumers to achieve quantified objectives with effectiveness, efficiency and satisfaction in a quantified context of use.

There are several standards for usability including the ISO 9241 and ISO 16982 standards, and the IEC 62366-1 standard (Applications of Usability Engineering to Medical Devices) from the International Electrotechnical Commission (IEC).

Usability, like quality, needs to be built into the software product rather than added later, and it needs to be considered from the earliest stages in the software development process. It requires an analysis of the user population and the tasks that they perform, as well as their knowledge and experience. The specification of the user and system requirements needs to include the usability requirements, as these are an integral part of the system.

There will often be a variety of different viewpoints to be considered, and this leads to multiple design solutions and an evaluation of these against the requirements. An iterative software development life cycle is generally employed, with active user involvement during the software development process. Prototyping is often employed to give the users a flavour of the proposed system and to get early user feedback on its usability. User acceptance testing (including usability testing) provides confidence that the software satisfies the usability, accessibility and quality expectations of the users (Table 7.2).

Table 7.2 Software development life cycle (including usability)

Phase	Description
Requirements	Interviews with the different categories of users
Prototype	Initial prototype developed and structured feedback given by users (usually via questionnaire)
Spiral design/ development	Design a little, code a little, test a little, formal review and user feedback prior to new spiral
Acceptance	Final acceptance testing by users

7.3.1 User-Centred Design

User-centred design (UCD) is a design process that is focused on the usability of and accessibility of the system to be developed, and it places the users at the centre of the software development process. The users are actively involved from the beginning of the project, and regular feedback is obtained from them at each stage of the process. UCD follows well-established techniques for analysis and design, and it is focused on understanding the characteristics of users and their needs (Table 7.3).

The UCD design activities focus on the user, including understanding the tasks that they perform, their needs and their experience. The users clarify what they want from the product and the environment in which the software will be used. The designers then determine how the users are currently performing their tasks, and what they like and dislike about the ways in which the tasks are currently done. This helps the designer to design a product that will be fit for purpose, that will satisfy the usability expectations of users, as well as being competitive in the market.

Table 7.3 UCD principles

Principle	Description
User understanding	The design is based on an explicit understanding of users, tasks and environments (i.e. who are the users?, what are their tasks and needs? and what is their experience?)
User involvement	The users are involved throughout the design and development (and user feedback shapes the design and development)
User evaluation	The design is driven and refined by user evaluation (and the user acceptance testing confirms that the usability and functional requirements are properly implemented)
Iterative development	The software development process is iterative, and the approach is to design and develop a little, get feedback from the user evaluation, modify accordingly and proceed to the next cycle in the iteration
Design	The design addresses the whole user experience
Multidisciplinary	The design team includes multidisciplinary skills

The software development team produces an initial version (or prototype) of the product, and the prototype has sufficient functionality to test some parts of the design. The design and development proceeds in cycles of modification, testing and a user review of the current version, until the software satisfies functional, usability and accessibility requirements. The approach is to design a little; code a little; test a little; evaluate and decide on whether to proceed with subsequent cycles.

A pre-release of the software may be created and sent to a restricted set of users for their evaluation, and the user feedback is then used to finalise the product prior to its actual release.

7.4   Review Questions

1. What is a text-based interface?
2. What is a graphical user interface?
3. Explain the importance of software usability.
4. Investigate the various usability standards such as ISO 9241 and ISO 16982.
5. Explain user-centred design.
6. Describe the evolution of human–computer interfaces.

7.5   Summary

Human–computer interaction is a branch of computer science that is concerned with the design, evaluation and implementation of interactive computing systems for human use. It is focused on the interfaces between people and computers, and has grown over the decades to include text-based interaction systems, graphical user interfaces and voice user interfaces.

The development of home computers from the mid-1970s meant that everyone in the world was now a potential computer user, and it was clear that there was a need to improve the usability of machines. Humans interact with computers in many ways, and so it is important to understand the interface between them to facilitate the interaction.

The early interaction between humans and computers was via batch processing with limited interaction between the operator and computer. These were followed by text-based interfaces (also known as a command line interface), where the system interaction (input and output) and navigation are text-based. One of the most well-known text-based operating systems was Microsoft's MS/DOS operating system for IBM compatible personal computers.

The graphical user interface is a human–computer interface that uses graphical icons, menus and windows to represent information and action to the user. They are intuitive and user-friendly, and were a revolution in human and computer interaction. They have made computers and electronic devices attractive to non-technical users, and are a major step forward from command-driven operating system.

The success of modern software systems is related to the usability of the software, and user-centred design has become a key paradigm in building usability in the software. It places the user at the centre of the software development process with active user involvement and evaluation employed.

Chapter 8
Introduction to Programming Languages

Key Topics

Generations of programming languages
Imperative Languages
ALGOL
Fortran and Cobol
Pascal and C
Object-oriented Languages
Java and C++
Functional Programming Languages
Logic Programming Languages
Syntax and Semantics

8.1 Introduction

Hardware is physical and can be seen and touched, whereas software is intangible and is an intellectual undertaking by a team of programmers. Software is written in a programming language, and hundreds of languages have been developed since the development of the early computers. Programming languages have evolved with the earliest languages using machine code to instruct the computer. The next development was the use of assembly languages to represent machine language instructions. These were then translated into machine code by an assembler. The next step was to develop high-level programming languages such as FORTRAN and COBOL. These were easier to use than assembly languages and machine code, and helped to improve quality and productivity.

A *first-generation* programming language (or 1GL) is a machine-level programming language that consists of 1s and 0s. The main advantage of these languages is execution speed as they may be directly executed on the computer, and

© Springer International Publishing AG, part of Springer Nature 2018

G. O'Regan, *World of Computing*,

https://doi.org/10.1007/978-3-319-75844-2_8

they do not require a compiler or assembler to convert from a high-level language or assembly language into the machine code.

However, writing a program in machine code is difficult and error prone, as it involves writing a stream of binary numbers. This made the programming language difficult to learn and difficult to correct should any errors occur. The programming instructions were entered through the front panel switches of the computer system, and adding new code was difficult. Further, the machine code was not portable as the machine language for one computer could differ significantly from that of another computer. Often, the program needed to be totally rewritten for the new computer. First-generation languages were used mainly on the early computers.

Second-generation languages, or 2GL, are low-level assembly languages that are specific to a computer and processor. However, assembly languages are easier to read than the first-generation machine code, and the assembler converts the assembly code into the actual machine code to run on the computer. The assembly language is specific to a particular processor family and environment, and it is therefore not portable. They require considerably more programming effort than high-level programming languages, and are more difficult to use for larger applications.

A program written in assembly language often needs to be rewritten for a different platform. However, since the assembly language is in the native language of the processor it has significant speed advantages over high-level languages. Second-generation languages are still used today, but high-level programming languages have generally replaced them.

The *third-generation languages*, or 3GL, include high-level programming languages such as Pascal, C or FORTRAN. These are general-purpose languages and have been applied to business, scientific and general applications. A program written in a high-level programming language is generally translated by the compiler[1] into the machine language of the target computer for execution. They are designed to be easier for a human to understand, and include features such as named variables, conditional statements, iterative statements, assignment statements and data structures. Early examples of third-generation languages are FORTRAN, ALGOL and COBOL, and later examples are C, C++ and Java. The advantages of these high-level languages are:

---

[1]This is true of code generated by native compilers. Other compilers may compile the source code to the object code of a Virtual Machine, and the translator module of the Virtual Machine translates the bytecode of the Virtual Machine to the corresponding native machine instruction. That is, the Virtual Machine translates each generalised machine instruction into a specific machine instruction (or instructions) that may then be executed by the processor on the target computer. A computer language such as C requires a separate compiler for each computer platform (i.e. computer and operating system). However, a language such as Java comes with a virtual machine for each platform. This allows the source code statements in these programs to be compiled just once, and they may then be executed on any platform.

- Ease of readability,
- Clearly defined syntax (and semantics[2]),
- Suitable for Business or Scientific applications,
- Machine independent,
- Portability to other platforms,
- Ease of debugging,
- Execution speed.

These languages are machine independent and may be compiled for different platforms. The early 3GLs were *procedural* in that they focus on how something is done rather than on what needs to be done. The later 3GLs were *object-oriented*[3] and the programming tasks were divided into objects. Objects may be employed to build larger programs, in a manner that is analogous to building a prefabricated building from its constituent parts. Java, C++ and Smalltalk are examples of modern object-oriented language.

High-level programming languages allow programmers to focus on problem solving rather than on the low-level details associated with assembly languages. They are easier to debug and to maintain than assembly languages.

Fourth-generation languages specify what needs to be done rather than how it should be done. They are designed to reduce programming effort and include report generators and form generators. Report generators take a description of the data format and the report that is to be created, and then automatically generate a program to produce the report. Form generators are used to generate programs to manage online interactions with the application system users. However, a disadvantage of 4GLs is that they are slow compared to compiled languages.

A *fifth-generation* programming language, or 5GL, is a programming language that is based around solving problems using constraints applied to the program, rather than using an algorithm written by the programmer. Fifth-generation languages are designed to make the computer (rather than the programmer) solve the problem. The programmer specifies the problem and the constraints to be satisfied, and is not concerned with the algorithm or implementation details. These languages are mainly used for research purposes especially in the field of artificial intelligence. Prolog is one of the best-known fifth-generation languages, and it is a logic programming language.

The task of deriving an efficient algorithm from a set of constraints for a particular problem is nontrivial and to date, this step has not been successfully automated. Fifth-generation languages are used mainly in academia.

---

[2]The study of programming language semantics commenced in the 1960s. It includes work done by Hoare on Axiomatic Semantics; work done by Gordon Plotkin on Operational Semantics; and work done by Scott and Strachey on Denotational Semantics.

[3]Norwegian Research originally developed object-oriented programming with their work on Simula 67 in the late 1960s.

8.2   Plankalkül

Plankalkül was developed by Konrad Zuse in 1946 and it is the earliest high-level programming language. It means "Plan" and "Kalkül": i.e. a Calculus of programs. It is a relatively modern language for such an old language, and there was no compiler available at the time of its creation. It was many years later before the first Plankalkül program was run, when the Free University of Berlin designed and developed a compiler for the language in 2000 (over 50 years after its conception).

The language employs data structures and Boolean algebra, and includes a mechanism to define more powerful data structures. Zuse demonstrated that the Plankalkül language could be used to solve scientific and engineering problems, and he wrote several example programs including programs for sorting lists and searching a list for a particular entry. The main features of Plankalkül are:

- A high-level language.
- Fundamental data types are arrays and tuples of arrays.
- While construct for iteration.
- Conditionals are addressed using guarded commands.
- There is no GOTO statement.
- Programs are non-recursive functions.
- Type of a variable is specified when it is used.

The main constructs of the language are variable assignment, arithmetical and logical operations, guarded commands and while loops. There are also some list and set processing functions.

8.3   Imperative Programming Languages

Imperative programming is a programming style that describes computation in terms of a program state, and statements that change the state. The term "*imperative*" is a command to carry out a specific instruction or action, and imperative programming consists of a set of commands to be executed on the computer, and is, therefore, concerned with *how* the program will be executed. The execution of an imperative command generally results in a change of state.

Imperative programming languages are quite distinct from *functional* and *logic programming languages*. Functional programming languages, like Miranda, have no global state, and programs consist of mathematical functions that have no side effects. In other words, there is no change of state, and the variable x will have the same value later in the program as it does earlier. Logic programming languages, like Prolog, define "*what*" is to be computed, rather than "*how*" the computation is to take place.

Most high-level programming languages are imperative languages, and assembly languages and machine code are also imperative languages. Imperative programs tend to be more difficult to reason about due to the change of state, as the variable x may have a different value later in the program. Functional programming and relational programming languages are mainly used in academia.

High-level imperative languages use program variables and employ commands such as assignment statements, conditional statements, iterative commands and calls to procedures. An assignment statement performs an operation on information located in memory, and stores the results in memory. Its effect is a change of the program state. A conditional statement allows a statement to be executed only if a specified condition is satisfied, whereas an iterative statement allows a statement (or a group of statements) to be executed multiple times while a specified condition is satisfied.

High-level imperative languages allow the evaluation of complex expressions such as arithmetic operations and function evaluations, and the resulting value of the expression is assigned to memory.

FORTRAN was developed in the mid-1950s, and it was one of the earliest programming languages. ALGOL was developed in the late 1950s and 1960s, and it became a popular language for the expression of algorithms. COBOL was designed in the late 1950s as a programming language for business use. George Kemeny and Thomas Kurtz designed the BASIC (beginner's all-purpose symbolic instruction code) programming language in the 1960s. Niklaus Wirth developed Pascal in the early 1970s. Denis Ritchie developed the C programming language at Bell Labs in the early 1970s.

The Ada programming language was developed for the US military in the early 1980s. Object-oriented languages include features to support objects, and Bjarne Stroustrup designed C++ in 1985 as an object-oriented extension of the C language. Sun Microsystems released Java in 1996.

8.3.1 FORTRAN and COBOL

FORTRAN (FORmula TRANslator) was the first high-level programming language to be implemented. John Backus at IBM developed it in the mid-1950s, and the first compiler was available in 1957. The language includes named variables, complex expressions and subprograms. It was designed for scientific and engineering applications, and remains the most important programming language for these domains. The main statements of the language include:

- Assignment Statements (using the = symbol),
- IF Statements,
- Goto Statements,
- DO Loops.

Fortran II was developed in 1958, and it introduced sub-programs and functions to support procedural (or imperative) programming. Each procedure (or subroutine) contains computational steps to be carried out when it is called (at any point) during program execution. This could include calls by other procedures or by itself. However, recursion was not allowed until Fortran 90. Fortran 2003 provides support for object-oriented programming.

The basic types supported in FORTRAN include Boolean, Integer and Real. Support for double precision and complex numbers was added later. The language includes relational operators for equality (.EQ.), less than (.LT.) and so on. FORTRAN is good at handling numbers and computation, and this is especially useful for mathematical and engineering problems. The following code (written in Fortran 77) gives a flavour of the language.

```
     PROGRAM HELLOWORLD
C    FORTRAN 77 SOURCE CODE COMMENTS FOR HELLOWORLD
     PRINT '(A)', 'HELLO WORLD'
     STOP
     END
```

FORTRAN remains a popular language for application such as climate modelling, simulations of the solar system, modelling the trajectories of artificial satellites and simulation of automobile crash dynamics.

It was initially weak at handling input and output, which was important to business computing. This led to the development of the COBOL programming language in the late 1950s.

The common business-oriented language (COBOL) was the first business programming language, and it was introduced in 1959. Grace Murray Hopper[4] (Fig. 8.1) and a group of computer professionals called the Conference on Data Systems Languages (CODASYL) designed it with the objective of improving the readability of software source code. It has an English-like syntax designed to make it easy to learn the language, and its only data types were numbers and strings of text, that may be grouped into arrays and records. The language is verbose:

'DIVIDE A BY B GIVING C REMAINDER D'.

COBOL was the first computer language, whose use was mandated by the US Department of Defense. The language remains in use today, and there is an object-oriented version of the language.

---

[4]Mary Hopper was a programmer on the Mark 1, Mark II and Mark III and UNIVAC 1 computers. She was the technical advisor to the CODASYL committee.

Fig. 8.1 Grace Murray and
UNIVAC

8.3.2 ALGOL

ALGOL (ALGOrithmic Language) is a family of imperative programming languages that was originally developed in the mid-1950s. It was later revised in ALGOL 60, and ALGOL 68, and the language was designed to address some of the problems in FORTRAN. ALGOL was not a widely used language, and this may have been due to the refusal of IBM to support ALGOL, and the dominance of IBM in the computing field.

A committee of American and European computer scientists designed the language, and ALGOL had a significant influence on later language design. ALGOL 60 (Naur 1960) was the most popular member of the family, and Edsger Dijkstra developed an early ALGOL 60 compiler. John Backus and Peter Naur developed a method for describing the syntax of the ALGOL 58 programming language, which is known as Backus–Naur form (or BNF).

ALGOL includes data structures and block structures. Block structures were designed to allow blocks of statements to be created (e.g. for procedures or functions). A variable defined within a block may be used within the block, but is out of scope outside of the block.

ALGOL 60 introduced two ways of passing parameters to sub-programs, and these are "*call by value*" and "*call by name*". The call by value parameter passing technique involves the evaluation of the arguments of a function or procedure before the function or procedure is entered. The values of the arguments are passed to the function or procedure, and any changes to the arguments within the called function or procedure have no effect on the actual arguments. The call by name parameter passing technique is the default parameter passing technique in ALGOL 60. It involves re-evaluating the actual parameter expression each time the formal parameter is read. Call by name is used today in C/C++ macro expansion.

ALGOL 60 includes conditional statements and iterative statements. It supports recursions: i.e. it allows a function or procedure to call itself. It includes:

- *Dynamic Arrays*: These are arrays in which the subscript range is specified by variables.
- *Reserved Words*: These are keywords that are not allowed to be used as identifiers by the programmer.
- *User-defined data types*: These allow the user to design their own data types.
- ALGOL uses bracketed statement blocks and it was the first language to use *begin end* pairs for delimiting blocks.

ALGOL was used mainly by researchers in the United States and Europe. There was a lack of interest in its adoption by commercial companies due to the absence of standard input and output facilities in its description. ALGOL 60 became the standard for the publication of algorithms, and it had a major influence on later language development.

ALGOL evolved during the 1960s but not in the right direction. The ALGOL 68 committee decided on a very complex design rather than the simple and elegant ALGOL 60 specification. Tony Hoare remarked that:

> "*ALGOL 60 was a great improvement on its successors*".

8.3.3   Pascal and C

Niklaus Wirth designed the Pascal programming language in the early 1970s. It is named after Blaise Pascal (a seventeenth-century French mathematician), and it was based on the ALGOL programming language. It was intended as a language to teach students structured programming.

Structured programming is concerned with rigorous techniques to design and develop programs, and there was an intense debate on correct approaches to software development in the late 1960s. Dijkstra argued against the use of the GOTO statement "GOTO Statement considered harmful" (Dijkstra 1968), and this influenced language design, and led to several languages that did not include the construct.

The Pascal language includes the conditional if statement, the iterative while, repeat and for statements, the assignment statement and the case statement (which is a generalised if statement). The statement in the body of the repeat statement is executed at least once, whereas the statement within the body of a while statement may never be executed.

The language has several reserved words (known as keywords) that have a special meaning, and these may not be used as program identifiers. The Pascal program that displays 'Hello World' is given by:

```
program HELLOWORLD (OUTPUT);
begin
     WRITELN ('Hello, World!')
end.
```

Pascal includes several simple data types such as Boolean, integer, character and reals, and it also has more advanced data types such as arrays, enumeration types, ordinal types and pointer data types. It allows complex data types to be constructed from existing data types, and types are introduced with the reserved word 'type'.

```
type
    c = record
            a: integer;
            b: char
        end;
```

Pascal includes a "pointer" data type, and this data type allows linked lists to be created by including a pointer type field in the record. The variable linklist is a pointer to the data type B in the example below where B is a record.

```
type
    BPTR = ^B;
    B    = record
            A: integer;
            C: BPTR
        end;
```

```
var
    linklist : BPTR;
```

Pascal is a block-structured language with programs structured into procedures and function blocks. These can be nested to any depth, and recursion is allowed. Each block has its own constants, types, variables and other procedures and functions, which are defined, within the scope of the block.

Pascal was criticised as being unsuitable for serious programming by Brian Kernighan and others (Kernighan 1981). Many of these deficiencies were addressed in later versions of the language. However, by then Denis Richie at Bell Labs had developed the C programming language, which became popular in the industry. C is a general-purpose and a systems programming language.

It was originally designed as a language to write the kernel for the UNIX operating system, which was novel as operating systems were traditionally written in assembly languages. The success of C in writing the UNIX kernel led to its use on several other operating systems such as Windows and Linux. It also influenced later language development such as C++, and it is one of the most commonly used systems programming languages. The language is described in detail in Kernighan and Ritchie (1978).

It provides high-level and low-level capabilities, and a C program that is written in ANSI C with portability in mind may be compiled for a very wide variety of computer platforms and operating systems with minimal changes to the source code. The C language is now available on a wide range of platforms.

C is a procedural programming language and includes conditional statements such as the "if statement", the "switch statement", iterative statements such as the "while" statement or "do" statement and the assignment statement.

- If Statement

```
if (A == B)
    A = A + 1;
else
    A = A - 1;⁵
```

- Assignment Statement

$i = i + 1;$

One of the first programs that people write in C is the Hello world program. This is given by:

```
main()
{
    printf("Hello, World\n");
}
```

It includes several predefined data types including integers and floating point numbers.

```
- int      (integer)
- long     (long integer)
- float    (floating point real)
- double (double precision real)
```

It allows more complex data types to be created using "structs", which are similar to records in Pascal. It allows the use of pointers to access memory locations, which allows the memory locations to be directly referenced and modified. For example, the result of the following is to assign 5 to the variable x.

```
int  x;
int *ptr_x;
x = 4;
ptr_x = &x;
*ptr_x = 5;
```

---

[5]The semi-colon in Pascal is used as a statement separator, whereas it is used as a   statement terminator in C.

C is a block-structured language, and a program is structured into functions (or blocks). Each function block contains variables and functions, and a function may call itself (i.e. recursion is allowed).

One key criticism of C is that it is very easy to make errors in C programs, and to thereby produce undesirable results. For example, one of the easiest mistakes to make is to accidentally write the assignment operator (=) for the equality operator (==). This totally changes the meaning of the original statement as can be seen below:

```
if (a == b)
      a ++;                          .... Program fragment A
else
      a--
```

```
if (a = b)
      a ++;                          .... Program fragment B
else
      a--
```

Both program fragments are syntactically correct and the intended meaning of a program is easily changed. The philosophy of C to allow statements to be written as concisely as possible, and this is potentially dangerous.[6] The use of pointers may lead to problems as uninitialised pointers may point anywhere in memory, and may, therefore, write anywhere in memory. Therefore, the effective use of C requires experienced programmers, well-documented source code and formal peer reviews of the source code by other developers.

8.4   Object-Oriented Languages

The traditional view of programming is that a program is a collection of functions, or a list of instructions to be performed on the computer. *Object-oriented programming* is a paradigm shift in programming, where a computer program is considered to be a collection of objects that act on each other. Each object may send and receive messages and process data. That is, each object may be viewed as an independent entity or actor with a distinct role or responsibility.

---

[6]It is very easy to write incomprehensible code in C and even a 1 line of C code can be incomprehensible. The maintenance of poorly written code is a challenge unless programmers follow good programming practice. This discipline needs to be enforced by formal reviews of the source code.

An object is a "black box" which sends and receives *messages*. A black box consists of *code* (computer instructions) and *data* (information which these instructions operate on). The traditional way of programming kept code and data separate. For example, functions and data structures in the C programming language are not connected. However, in the object-oriented world, code and data are Merged into a single indivisible thing called an *object*.

The reason that an object is called a black box is that the user of an object never needs to look inside the box, since all communication to it is done via messages. Messages define the *interface* to the object. Everything an object can do is represented by its message interface. Therefore, there is no need to know anything about what is in the black box (or object) to use it. The access to an object is only through its messages, while keeping the internal details private. This is called *information hiding*[7] and is due to work by Parnas in the early 1970s.

The origins of object-oriented programming go back to the invention of Simula 67 at the Norwegian Computing Research Centre[8] in the late 1960s. Simula 67 introduced the notion of a class and instances of a class[9], and it influenced later languages such as Smalltalk developed at Xerox PARC in the mid-1970s. Xerox introduced the term '*Object-oriented programming*' for the use of objects and messages as the basis for computation. Most modern programming languages support object-oriented programming (e.g. Java and C++), and object-oriented features added to many existing languages such as BASIC, FORTRAN and Ada. The main features of object-oriented languages are described in Table 8.1.

Object-oriented programming has become popular in large-scale software development, and it became the dominant paradigm in programming from the early 1990s. Its proponents argue that it is easier to learn, and simpler to develop and maintain such programs. Its growth in popularity was helped by the rise in popularity of graphical user interfaces (GUI), which is well suited to object-oriented programming. The C++ programming language has become popular, and it is an object-oriented extension of the C programming language.

8.4.1 C++ and Java

Bjarne Stroustrup developed the C++ programming language in 1983 as an object-oriented extension of the C programming language. It was designed to use the power of object-oriented programming, and to maintain the speed and

---

[7]Information hiding is a key contribution by Parnas to computer science. He has also done work on mathematical approaches to software quality using tabular expressions (O'Regan 2017).

[8]The inventors of Simula 67 were Ole-Johan Dahl and Kristen Nygaard.

[9]Dahl and Nygaard were working on ship simulations and were attempting to address the huge number of combinations of different attributes from different types of ships. Their insight was to group the different types of ships into different classes of objects, with each class of objects being responsible for defining its own data and behaviour

Table 8.1   Object-oriented paradigm

Feature	Description
Class	A class defines the abstract characteristics of a thing, including its attributes (or properties), and its behaviours (or methods). The members of a class are termed objects
Object	An object is a particular instance of a class with its own set of attributes. The set of values of the attributes of an object is called its state
Method	The methods associated with a class represent the behaviours of the objects in the class
Message passing	Message passing is the process by which an object sends data to another object, or asks the other object to invoke a method
Inheritance	A class may have subclasses (or children classes) that are more specialised versions of the class. A subclass inherits the attributes and methods of the parent class. This allows the programmer to create new classes from existing classes. The derived classes inherit the methods and data structures of the parent class
Encapsulation (information hiding)	One fundamental principle of the object-oriented world is encapsulation (or information hiding). The internals of an object are kept private to the object, and may not be accessed from outside the object. That is, encapsulation hides the details of how a particular class is implemented, and it requires a clearly specified interface around the services provided
Abstraction	Abstraction simplifies complexity by modelling classes and removing all unnecessary detail. All essential detail is represented, and non-essential information is ignored
Polymorphism	Polymorphism is behaviour that varies depending on the class in which the behaviour is invoked. Two or more classes may react differently to the same message. The same name is given to methods in different subclasses: i.e. one interface, and multiple methods

portability of C. It provides a significant extension of Cs capabilities, but it does not force the programmer to use the object-oriented features of the language.

A key difference between C++ and C is the concept of a class. A *class* is an extension to the C concept of a structure, where the main difference is that while a C data structure can hold only data, a C++ class may hold both data and functions. An *object* is an instantiation of a class: i.e. the class is essentially the type, whereas the object is essentially a variable of that type. Classes are defined in C++ by using the keyword class:

```
class class_name
{
        access_specifier_1:
            member1;
        access_specifier_2:
```

```
            member2;
...
}
```

The members may be either data or function declarations, and an access specifier is included to specify the access rights for each member (e.g. private, public or protected). Private members of a class are accessible only by other members of the same class, public members are accessible from anywhere where the object is visible, protected are accessible by other members of same class and also from members of their derived classes. An example of a class in C++ is the definition of the class rectangle:

```
class CRectangle
{
    int x, y;
        public:
            void set_values (int, int);
            int area (void);
} rect;
```

Java is an object-oriented programming language developed by James Gosling and others at Sun Microsystems in the early 1990s. C and C++ influenced the syntax of the language, and Java was designed with portability in mind. The objective is for a program to be written once and executed anywhere. Platform independence is achieved by compiling the Java code into Java bytecode, which are simplified machine instructions specific to the Java platform.

This code is then run on a Java virtual machine (JVM) that interprets and executes the Java bytecode. The JVM is specific to the native code on the host hardware. The problem with interpreting bytecode is that it is slow compared to traditional compilation. However, Java has a number of techniques to address this including just in time compilation and dynamic recompilation. Java also provides automatic garbage collection. This is a very useful feature as it protects programmers who forget to deallocate memory (thereby causing memory leaks).

Java is a proprietary standard that is controlled through the Java Community Process. Sun Microsystems makes most of its Java implementations available without charge. The following is an example of the Hello World program written in Java.

```
class HelloWorld
{
        public static void main (String args[])
        {
                System.out.println ("Hello World!");
        }
}
```

8.5   Functional Programming Languages

Functional programming is quite distinct from imperative programming in that it involves the evaluation of mathematical functions. Imperative programming involves the execution of sequential (or iterative) commands that change the state. For example, the assignment statement alters the value of a variable, and the value of a given variable x may change during program execution.

There is no change of state in functional programs, and the fact that the value of x will always be the same makes it easier to reason about functional programs than imperative programs. Functional programming languages provide *referential transparency*: i.e. equals may be substituted for equals, and if two expressions have equal values, then one can be substituted for the other in any larger expression without affecting the result of the computation.

Functional programming languages use higher order functions,[10] recursion, lazy and eager evaluation, monads,[11] and Hindley–Milner-type inference systems.[12] These languages are mainly used in academia, but there has been some industrial use, including the use of Erlang for concurrent applications in industry. Alonzo Church developed Lambda Calculus in the 1930s, and it provides an abstract framework for describing mathematical functions and their evaluation. It provides the foundation for functional programming languages, and Church employed lambda Calculus to prove that there is no solution to the decision problem for first-order arithmetic in 1936.

Lambda Calculus uses transformation rules, and one of these rules is variable substitution. The original Calculus developed by Church was untyped, but typed lambda calculi have since been developed. Any computable function can be expressed and evaluated using lambda Calculus, but there is no general algorithm to determine whether two arbitrary lambda Calculus expressions are equivalent. Lambda Calculus influenced functional programming languages such as Lisp, ML and Haskell.

Functional programming uses the notion of higher order functions. Higher order takes other functions as arguments, and may return functions as results. The derivative function $\frac{d}{dx} f(x) = f'(x)$ is a higher order function that takes a function as an argument and returns a function as a result. For example, the derivative of the function $Sin(x)$ is given by $Cos(x)$. Higher order functions allow currying which is a technique developed by Schönfinkel. It allows a function with several arguments to be applied to each of its arguments one at a time, with each application returning a

---

[10]Higher order functions are functions that take functions as arguments or return a function as a result. They are known as operators (or functionals) in mathematics.

[11]Monads are used in functional programming to express input and output operations without introducing side effects. The Haskell functional programming language makes use of uses this feature.

[12]This is the most common algorithm used to perform type inference, which is concerned with determining the type of the value derived from the eventual evaluation of an expression.

new (higher order) function that accepts the next argument. This allows a function of n-arguments to be treated as n applications of a function with 1-argument.

John McCarthy developed LISP at MIT in the late 1950s, which includes many of the features found in modern functional programming languages.[13] Scheme built upon the ideas in LISP, and Kenneth Iverson developed APL[14] in the early 1960s. APL influenced Backus's FP programming language, and Robin Milner designed the ML programming language in the early 1970s. David Turner developed Miranda in the mid-1980s, and it influenced the Haskell programming language developed by Philip Wadler and others in the late 1980s/early 1990s.

8.5.1   Miranda

Miranda was developed by David Turner at the University of Kent in the mid-1980s (Turner 1985). It is a non-strict functional programming language: i.e. the arguments to a function are not evaluated until they are required within the function being called. This is also known as *lazy evaluation*, and one of its key advantages is that it allows a potentially infinite data structure to be passed as an argument to a function. Miranda is a pure functional language in that there are no side effect features in the language. The language has been used for:

- Rapid prototyping
- Specification language
- Teaching Language

A Miranda program is a collection of equations that define various functions and data structures. It is a strongly typed language with a terse notation.

$$z = \text{sqr}\, p \,/\, \text{sqr}\, q$$
$$\text{sqr}\, k = k * k$$
$$p = a + b$$
$$q = a - b$$
$$a = 10$$
$$b = 5$$

The scope of a formal parameter (e.g. the parameter k above in the function sqr) is limited to the definition of the function in which it occurs.

One of the most common data structures used in Miranda is the list. The empty list is denoted by [ ], and an example of a list of integers is given by [1, 3, 4, 8]. Lists may be appended to by using the "++" operator. For example:

---

[13]Lisp is a multi-paradigm language rather than a functional programming language.

[14]Iverson received the Turing Award in 1979 for his contributions to programming language and mathematical notation. The title of his Turing award paper was "Notation as a tool of thought".

$$[1,3,5] + +[2,4] \text{ is } [1,3,5,2,4]$$

The length of a list is given by the "#" operator:

$$\#[1,3] = 2$$

The infix operator ":" is employed to prefix an element to the front of a list. For example:

$$5 : [2,4,6] \text{ is equal to } [5,2,4,6]$$

The subscript operator "!" is employed for subscripting: For example:

$$\text{Nums} = [5,2,4,6] \quad \text{then Nums!0 is 5}$$

The elements of a list are required to be of the same type. A sequence of elements that contains mixed types is called a tuple. A tuple is written as follows:
Employee = ("Holmes", "221B Baker St. London", 50, "Detective")
A tuple is similar to a record in Pascal, whereas lists are similar to arrays. Tuples cannot be subscripted but their elements may be extracted by pattern matching. Pattern matching is illustrated by the well-known example of the factorial function:

$$\text{fac } 0 = 1$$
$$\text{fac } (n+1) = (n+1) * \text{ fac n}$$

The definition of the factorial function uses two equations, distinguished by using different patterns in the formal parameters. Another example of pattern matching is the reverse function on lists:

$$\text{reverse } [\,] = [\,]$$
$$\text{reverse}(a : x) = \text{reverse } x ++ [a]$$

Miranda is a higher order language, and it allows functions to be passed as parameters and returned as results. Currying is allowed and this allows a function of n-arguments to be treated as n applications of a function with 1-argument. Function application is left associative: i.e. $f x y$ means $(f x) y$. That is, the result of applying the function f to x is a function, and this function is then applied to y. Every function with two or more arguments in Miranda is a higher order function.

8.5.2   Lambda Calculus

Lambda Calculus (λ-Calculus) was designed by Alonzo Church in the 1930s to study computability. It is a formal system that may be used to study function definition, function application, parameter passing and recursion. Any computable function may be expressed and evaluated using lambda Calculus.

Lambda Calculus is equivalent to the abstract Turing machine formalism in that they compute the same set of functions. However, lambda Calculus emphasises the use of transformation rules, whereas Turing machines are concerned with computability on primitive machines. Lambda Calculus consists of a small set of rules:

Alpha-conversion rule (α-conversion)[15]
Beta-reduction rule (β-reduction)[16]
Eta-conversion (η-conversion)[17]

Every expression in the λ-Calculus stands for a function with a single argument. The argument of the function is itself a function with a single argument and so on. The definition of a function is anonymous in the Calculus. For example, the function that adds one to its argument is usually defined as $f(x) = x + 1$. However, in λ-Calculus the function is defined as:

$$\lambda x \cdot x + 1$$

The name of the formal argument x is irrelevant and an equivalent definition of the function is $\lambda z \cdot z + 1$. The evaluation of a function f with respect to an argument (e.g. 3) is usually expressed by $f(3)$. In λ-Calculus this would be written as ($\lambda x \cdot x + 1$) 3, and this evaluates to $3 + 1 = 4$. Function application is left associative: i.e. $f\, x\, y = (f\, x)\, y$. A function of two variables is expressed in lambda Calculus as a function of one argument, which returns a function of one argument. This is known as currying and has been discussed earlier. For example, the function $f(x, y) = x + y$ is written as $\lambda x \cdot \lambda y \cdot x + y$. This is often abbreviated to $\lambda x\, y \cdot x + y$.

λ-Calculus is a simple mathematical system and its syntax is defined as follows:

```
<exp> ::= <identifier>          |
          λ <identifier>.<exp>  | --abstraction
          <exp> <exp>           | --application
          ( <exp> )

          -- Syntax of Lambda Calculus --
```

λ-Calculus's four lines of syntax plus *conversion* rules, are sufficient to define Booleans, integers, data structures and computations on them. It inspired LISP and modern functional programming languages.

---

[15]This essentially expresses that the names of bound variables are unimportant.

[16]This essentially expresses the idea of function application.

[17]This essentially expresses the idea that two functions are equal if and only if they give the same results for all arguments.

8.6   Logic Programming Languages

Logic programming languages describe what is to be done, rather than how it should be done. These languages are concerned with the statement of the problem to be solved, rather than how the problem will be solved.

These languages use mathematical logic as a tool in the statement of the problem definition. Logic is a useful tool in developing a body of knowledge (or theory), and it allows rigorous deduction of further truths from the existing set of truths. The theory is built up from a small set of axioms or postulates and rules of inference to derive further truths logically.

The objective of logic programming is to employ mathematical logic to assist with computer programming. Many problems are naturally expressed as a theory, and the statement of a problem to be solved is often equivalent to determining if a new hypothesis is consistent with an existing theory. Logic provides a rigorous way to determine this, as it includes a rigorous process for conducting proof.

Computation in logic programming is essentially logical deduction, and logic programming languages use first-order[18] predicate Calculus. It employs theorem proving to derive the desired truth from an initial set of axioms. These proofs are constructive[19] in the sense that an actual object that satisfies the constraints is produced, rather than a reliance on a theoretical existence theorem. Logic programming specifies the objects, the relationships between them and the constraints that must be satisfied for the problem. It specifies:

-   The set of objects involved in the computation.
-   The relationships that hold between the objects.
-   The constraints that must be satisfied for the problem.

The language interpreter then decides how to satisfy the constraints. Artificial intelligence influenced the development of logic programming, and John McCarthy[20] demonstrated that mathematical logic could be used for expressing knowledge. The first logic programming language was Planner developed by Carl Hewitt at MIT in 1969. It uses a procedural approach for knowledge representation rather than McCarthy's declarative approach.

The best-known logic programming languages is Prolog, which was developed in the early 1970s by Alain Colmerauer and Robert Kowalski. It stands for *pro*gramming in *log*ic. It is a goal-oriented language that is based on predicate

---

[18]First-order logic allows quantification over objects but not functions or relations. Higher order logics allow quantification of functions and relations.

[19]For example, the constructive proof of the statement $\exists x$ such that $x = \sqrt{4}$ (i.e. there is an x such that x is the square root of 4) provides more than a proof of existence, and an actual object satisfying the existence criteria is explicitly produced (i.e. that $x = 2$ or $x - -2$).

[20]John McCarthy received the Turing Award in 1971 for his contributions to artificial intelligence. He also developed the programming language LISP.

logic. Prolog became an ISO standard in 1995. The language attempts to solve a goal by tackling the sub-goals that the goal consists of:

$$\text{goal: - subgoal}_1, \ldots, \text{subgoal}_n.$$

That is, in order to prove a particular goal it is sufficient to prove subgoal$_1$ through subgoal$_n$. Each line of a Prolog program consists of a rule or a fact, and the language specifies what exists rather than how. The following program fragment has one rule and two facts:

grandmother(G,S) :- parent(P,S), mother(G,P).mother(Sarah, isaac).parent(isaac, Jacob).

The first line in the program fragment is a rule that states that G is the grand-mother of S, if there is a parent P of S and G is the mother of P. The next two statements are facts stating that Isaac is a parent of Jacob, and that Sarah is the mother of isaac. A goal clause is true if all of its subclauses are true:

$$\text{goalclause}\left(V_g\right) : -\text{clause}_1(V_1), \ldots, \text{clause}_m(V_m)$$

A Horn clause consists of a goal clause and a set of clauses that must be proven separately. Prolog finds solutions by *unification*: i.e. by binding a variable to a value. For an implication to succeed, all goal variables Vg on the left side of:- must find a solution by binding variables from the clauses which are activated on the right side. When all clauses are examined and all variables in Vg are bound, the goal succeeds. But if a variable cannot be bound for a given clause, then that clause fails. Following the failure, Prolog *backtracks*, and this involves going back to the left to previous clauses to continue trying to unify with alternative bindings. Backtracking gives Prolog the ability to find multiple solutions to a given query or goal.

Most logic programming languages use a simple searching strategy to consider alternatives:

If a goal succeeds and there are more goals to achieve, then remember any untried alternatives and go on to the next goal.
If a goal is achieved and there are no more goals to achieve then stop with success.
If a goal fails and there are alternative ways to solve it then try the next one.
If a goal fails and there are no alternate ways to solve it, and there is a previous goal, then go back to the previous goal.
If a goal fails and there are no alternate ways to solve it, and no previous goal, then stop with failure.

Constraint programming is a programming paradigm, where relations between variables can be stated in the form of constraints. Constraints specify the properties of the solution, and it differs from imperative programming in that the sequence of steps to execute to establish the solution is not specified.

8.7 Syntax and Semantics

There are two key parts to any programming language namely its syntax and semantics. The syntax is the grammar of the language, and a program needs to be syntactically correct with respect to its grammar. The semantics of the language is deeper, and determines the meaning of what has been written by the programmer. The semantics of a language determines what a syntactically valid program will compute. A programming language is therefore given by:

$$Programming\ Language = Syntax + Semantics$$

The theory of the syntax of programming languages is well established, and Chomsky[21] defined a hierarchy of grammars (regular, context-free, context-sensitive). Backus–Naur form[22] (BNF) is often employed to specify the grammar of languages, which may be input into a parser to determine whether the program is syntactically correct. A BNF specification consists of a set of rules such as:

$$<symbol> ::= expression\ with\ symbols >$$

Where <symbol> is a *non-terminal* and the expression consists of sequences of symbols and/or sequences separated by the vertical bar "|" which indicates a choice. Symbols that never appear on a left side are called *terminals*. The partial definition of the syntax of various statements in a programming language is given below:

```
< loop statement > :: =  < while loop >  |  < for loop>
<while loop >  :: = while (< condition >)  < statement>
<for loop >  :: = for (< expression >)  < statement>
<statement >  :: =  < assignment statement >  |  < loop statement>
<assignment statement >  :: =  < variable >  : =  < expression>
```

The example above includes various non-terminals (<loop statement>, <while loop>, <for loop>, <condition>, <expression>, <statement>, <assignment statement>, and <variable>). The terminals include "while", "for", ":=" "("and")". The production rules for <condition> and <expression> are not included.

There are various types of grammars such as regular grammars, context-free grammars, and context-sensitive grammars. A parser translates the grammar of a

---

[21]Chomsky made important contributions to linguistics and the theory of grammars. He is more widely known today as a critic of United States foreign policy.

[22]Backus–Naur form is named after John Backus and Peter Naur. It was created as part of the design of Algol 60, and used to define the syntax rules of the language.

Table 8.2 Programming language semantics

Approach	Description
Axiomatic semantics	Axiomatic semantics involves giving meaning to phrases of the language with logical axioms. This approach is based on mathematical logic, and it employs pre- and post-condition assertions to specify what happens when the statement executes. The relationship between the initial assertion and the final assertion essentially gives the semantics of the code
Operational semantics	The operational semantics for a programming language was developed by Plotkin (1981). It describes how a valid program is interpreted by a sequence of computational steps. An abstract machine (SECD machine) may be defined to give meaning to phrases, by describing the transitions they induce on states of the machine. A precise mathematical interpreter (such as the lambda Calculus) may also give the semantics
Denotational semantics	Denotational semantics (originally called mathematical semantics) provides meaning to programs in terms of mathematical objects such as integers, tuples and functions. Each phrase in the language is translated into a mathematical object that is the *denotation* of the phrase. Christopher Strachey and Dana Scott developed it in the mid-1960s

language into a parse table, and each type of grammar has its own parsing algorithm to determine whether a particular program is syntactically correct with respect to its grammar.

8.7.1 Programming Language Semantics

The formal semantics of a programming language is concerned with the meaning of programs. A program is written according to the rules of its grammar (syntax), and the compiler then checks that it is syntactically correct, and if so, it generates the equivalent machine code.[23]

The compiler must preserve the semantics of the language, and the syntax of the language gives no information as to the meaning of a program. It is possible to write syntactically correct programs that behave in quite a different way from the intentions of the programmer.

The formal semantics of a language is given by a mathematical model, which describes the possible computations described by the language. The three main approaches to programming language semantic are axiomatic semantics, operational semantics and denotational semantics. A short summary of each approach is described in Table 8.2.

---

[23]Of course, what the programmer has written may not be what the programmer had intended.

8.8   Review Questions

1. Describe the five generations of programming languages.
2. Explain the difference between machine code and assembly languages.
3. What are the key features of Fortran and COBOL.
4. Describe the key features of Pascal and C.
5. What are the key features of object-oriented languages.
6. Explain the differences between imperative programming languages and functional programming languages.
7. What are the key features of logic programming languages?
8. What is the difference between syntax and semantics?
9. Explain the main approaches to programming language semantics.

8.9   Summary

This chapter considered the evolution of programming languages from the older machine languages, to the low-level assembly languages, to high-level programming languages and object-oriented languages, and to functional and logic programming languages. The syntax and semantics of programming languages were briefly discussed.

The advantages of machine languages are execution speed and efficiency. However, it is difficult to write programs in these languages, as the program involves a stream of binary numbers. Further, these languages are not portable, as the machine language for one computer may differ significantly from the machine language of another.

The second-generation languages are low-level assembly languages that are specific to a particular computer and processor. These are easier to write and understand, but they must be converted into the actual machine code to run on the computer. They are specific to a processor family and environment and are not portable. However, their advantages are execution speed, as the assembly language is the native language of the processor.

The third-generation languages are high-level programming languages, and have been applied to business, scientific and general applications. They are designed to be easier to understand, and to allow the programmer to focus on problem-solving. Their advantages include ease of readability, portability and ease of debugging and maintenance. The early 3GLs were procedure-oriented and the later 3GLs were object-oriented.

Fourth-generation languages consist of statements similar to human language, and are often used in database programming. They specify what needs to be done rather than how it should be done, and they have been used as report generators and form generators.

Fifth-generation programming languages or 5GLs, are programming languages that are based around solving problems using logic programming or applying constraints to the program. They are designed to make the computer (rather than the programmer) solve the problem. The programmer only needs to be concerned with the specification of the problem and the constraints to be satisfied, and does not need to be concerned with the algorithm or implementation details.

Chapter 9
Overview of Software Engineering

Key Topics

Standish Chaos Report
Software Lifecycles
Waterfall Model
Spiral Model
Rational Unified Process
Agile Development
Software Inspections
Software Testing
Project Management
CMMI

9.1 Introduction

The approach to software development in the 1950s and 1960s has been described as the '*Mongolian Hordes Approach*' by Brooks (1975)[1]. The 'method' or lack of method was applied to projects that were running late, and it involved adding many inexperienced programmers to the project, with the expectation that this would allow the project schedule to be recovered. However, this approach was deeply flawed as it led to inexperienced programmers with inadequate knowledge of the project attempting to solve problems, and they inevitably required significant time from the other project team members resulting in further delays.

This led to the project being delivered even later, as well as subsequent problems with quality (i.e. the approach of throwing people at a problem does not work). The philosophy of software development back in the 1950/60s was characterised by:

---

[1]The 'Mongolian Hordes' management myth is the belief that adding more programmers to a software project that is running late will allow catch-up. In fact, as Brooks says adding people to a late software project actually makes it later.

© Springer International Publishing AG, part of Springer Nature 2018
G. O'Regan, *World of Computing*,
https://doi.org/10.1007/978-3-319-75844-2_9

The completed code will always be full of defects.

The coding should be finished quickly to correct these defects.

Design as you code approach.

This philosophy accepted defeat in software development, and suggested that irrespective of a solid engineering approach, that the completed software would always contain lots of defects. It, therefore, made sense to code as quickly as possible, and to then identify the defects that were present in order to correct them as quickly as possible.

In the late 1960s, it was clear that the existing approaches to software development were deeply flawed, and that there was an urgent need for change. The NATO Science Committee organised two famous conferences to discuss critical issues in software development (Buxton 1975). The first conference was held at Garmisch, Germany, in 1968, and it was followed by a second conference in Rome in 1969. Over 50 people from 11 countries attended the Garmisch conference, including Edsger Dijkstra, who did important theoretical work on formal specification and verification. The NATO conferences highlighted the problems that existed in the software sector in the late 1960s, and the term '*software crisis*' was coined to refer to these. There were problems with budget and schedule overruns, as well as problems with the quality and reliability of the delivered software.

The conference led to the birth of *software engineering* as a discipline, and the realisation that programming is quite distinct from science and mathematics. Programmers are like engineers in that they build software products, and so they need an education in traditional engineering as well as on the latest technologies. The education of a classical engineer includes product design and mathematics. However, often computer science education places an emphasis on the latest technologies, rather than on the essential engineering foundations for designing and building high-quality products that are safe for the public to use.

Programmers, therefore, need to learn the key engineering skills to enable them to build products that are safe for the public to use. These engineering skills include a solid foundation on design and on the mathematics required for building safe software products. Mathematics plays a key role in classical engineering, and it has a role to play in some situations (especially in the safety-critical and security-critical fields) in developing high-quality software products. Several mathematical approaches that may assist software engineers are described in (O'Regan 2017b).

There are parallels between the software crisis of the late 1960s, and the crisis with bridge construction in the nineteenth century. Several bridges collapsed or were delivered late or overbudget in the nineteenth century, because people involved in their design and construction did not have the required engineering knowledge. This led to bridges that were poorly designed and constructed, leading to their collapse and loss of life, as well as endangering the lives of the public.

This led to legislation requiring engineers to be licensed by the Professional Engineering Association prior to practicing as engineers. This organisation specified a core body of knowledge that the engineer is required to possess, and the licensing body verifies that the engineer has the required qualifications and

Fig. 9.1 Standish report—results of 1995 and 2009 survey

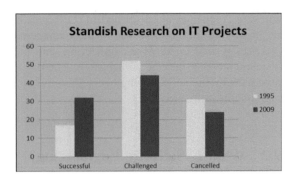

experience. This helps to ensure that only personnel competent to design and build products actually do so. Engineers have a professional responsibility to ensure that the products are properly designed and built, and are safe for the public to use.

The Standish group has conducted research (Fig. 9.1) on the extent of problems with IT projects since the mid-1990s. These studies were conducted in the United States, but there is no reason to believe that European or Asian companies perform any better. The results indicate serious problems with on time delivery of projects, and projects being cancelled prior to completion.[2] However, the comparison between 1995 and 2009 suggests that there have been some improvements with a greater percentage of projects being delivered successfully, and a reduction in the percentage of projects being cancelled.

Fred Brooks argues that software is inherently complex, and that there is no *silver bullet* that will resolve all problems associated with software development such as schedule or budget overruns (Brooks 1975; Brooks 1986). Poor software quality can lead to defects in the software that may adversely impact the customer, or even cause loss of life. It is, therefore, essential that software development organisations place sufficient emphasis on quality throughout the software development process.

The Y2K problem was caused by a two-digit representation of dates, and it required major rework to enable legacy software to function for the new millennium. Clearly, well-designed programs would have hidden the representation of the date, which would have led to minimal changes for year 2000 compliance. Instead, companies spent vast sums of money to rectify the problem.

The quality of software produced by some companies is impressive.[3] These companies employ mature software processes, and are committed to continuous improvement. There is a lot of industrial interest in software process maturity models for software organisations, and various approaches to assess and mature

---

[2]These are IT projects covering diverse sectors including banking, telecommunications, etc., rather than pure software companies. Software companies following maturity frameworks such as the CMMI generally achieve more consistent results.

[3]I recall projects at Motorola that regularly achieved 5.6σ-quality in a L4 CMM environment (i.e. approx. 20 defects per million lines of code. This represents very high quality).

software companies are described in (O'Regan 2010, 2014).[4] These models focus on improving the effectiveness of the management, engineering and organisation practices related to software engineering, and in introducing best practice in software engineering into the organisation. The disciplined use of the mature software processes by the software engineers plays a key role in the consistent delivery of high-quality software.

9.2   What Is Software Engineering?

Software engineering involves the multi-person construction of multi-version programs. The IEEE 610.12 definition of Software Engineering is:

> Software engineering is the application of a systematic, disciplined, quantifiable approach to the development, operation, and maintenance of software; that is, the application of engineering to software, and the study of such approaches.

Software engineering includes:

1. Methodologies to design, develop and test software to meet customers' needs.
2. Software is engineered. That is, the software products are properly designed, developed and tested in accordance with engineering principles.
3. Quality and safety are properly addressed.
4. Mathematics may be employed to assist with the design and verification of software products. The level of mathematics employed will depend on the *safety-critical* nature of the product. Systematic peer reviews and rigorous testing will often be sufficient to build quality into the software, with heavy *mathematical techniques reserved for safety- and security-critical software.*
5. Sound project management and quality management practices are employed.
6. Support and maintenance of the software are properly addressed.

Software engineering is not just programming. It requires the engineer to state precisely the requirements that the software product is to satisfy, and then to produce designs that will meet these requirements. The project needs to be planned and delivered on time and budget. The requirements must provide a precise description of the problem to be solved: i.e. *it should be evident from the requirements what is and what is not required.*

The requirements need to be rigorously reviewed to ensure that they are clear and unambiguous, and reflect the customer's needs. The next step is then to create

---

[4]Approaches such as the CMM or SPICE (ISO 15504) focus mainly on the management and organisational practices required in software engineering. The emphasis is on defining software processes that are fit for the purpose and consistently following them. The process maturity models focus on what needs to be done rather how it should be done. This gives the organisation the freedom to choose the appropriate implementation to meet its needs. The models provide useful information on practices to consider in the implementation.

the design that will solve the problem, and it is essential to validate the correctness of the design. Next, the software code to implement the design is written, and peer reviews and software testing are employed to verify and validate the correctness of the software.

The verification and validation of the design are rigorously performed for safety-critical systems, where it may be appropriate to employ mathematical techniques. However, it will usually be sufficient to employ peer reviews or software inspections for verification and validation, as these methodologies provide a high degree of rigour. This may include approaches such as Fagan inspections (Fagan 1976), Gilb inspections (Gilb and Graham 1994), or Prince 2's approach to quality reviews (Office of Government Commerce 2004).

The term *'engineer'* is a title that is awarded on merit in classical engineering. It is generally applied only to people who have attained the necessary education and competence to be called engineers, and who base their practice on classical engineering principles. The title places responsibilities on its holder to behave professionally and ethically. Often in computer science, the term *'software engineer'* is employed loosely to refer to anyone who builds things, rather than to an individual with a core set of knowledge, experience and competence.

Several computer scientists (such as Parnas[5]) have argued that computer scientists should be educated as engineers to enable them to apply appropriate scientific principles to their work. They argue that computer scientists should receive a solid foundation in mathematics and design, to enable them to have the professional competence to perform as engineers in building high-quality products that are safe for the public to use. The use of mathematics is an integral part of the engineer's work in other engineering disciplines, and so the *software engineer* should be able to use mathematics to assist with modelling and understanding the behaviour or properties of the proposed software system.

Software engineers need education[6] on specification, design, turning designs into programs, software inspections and testing. The education should enable the software engineer to produce well-structured programs that are fit for purpose.

Parnas has argued that software engineers have responsibilities as professional engineers.[7] They are responsible for designing and implementing high-quality and

---

[5]Parnas has made important contributions to computer science. He advocates a solid engineering approach with the extensive use of classical mathematical techniques in software development. He also introduced information hiding in the 1970s, which is now a part of object-oriented design.

[6]Software Companies that are following approaches such as the CMM or ISO 9001 consider the education and qualification of staff prior to assigning staff to performing specific roles. The appropriate qualifications and experience for the specific role are considered prior to appointing a person to carry out the role. Many companies are committed to the education and continuous development of their staff.

[7]The ancient Babylonians used the concept of accountability in the code of laws known as the Hammurabi Code, c. 1750 B.C. It included a law that stated that if a house collapsed and killed the owner then the builder of the house would be executed.

reliable software that is safe to use. They are also accountable for their decisions and actions,[8] and have a responsibility to object to decisions that violate professional standards. Engineers are required to behave professionally and ethically with their clients. The membership of the professional engineering body requires the member to adhere to the code of ethics[9] of the profession. Engineers in other professions are licensed, and therefore Parnas argues that a similar licensing approach be adopted for professional software engineers[10] to provide confidence that they are competent for the assignment. Professional software engineers are required to follow best practice in software engineering and the defined software processes.[11] Chapter 16 discusses ethics and professional responsibility.

Many software companies invest heavily in training, as the education and knowledge of its staff are essential to delivering high-quality products and services. Further, as the computer sector is rapidly changing, employees need to regularly reskill during their careers. Employees receive professional training related to the roles that they are performing, and the fact that the employees are professionally qualified increases confidence in the ability of the company to deliver high-quality products and services. A company that pays little attention to the competence and continuous development of its staff will obtain poor results, and suffer a loss of reputation and market share.

---

[8]However, it is unlikely that an individual programmer would be subject to litigation where a program failure causes damage or loss of life. A comprehensive disclaimer of responsibility for problems rather than a guarantee of quality accompanies most software products. Software engineering is a team-based activity involving many engineers in various parts of the project, and it would be potentially difficult for an outside party to prove that the cause of failure was due to the professional negligence of an individual software engineer, as there are many others involved in the process such as reviewers, testers and the entire project team that developed the software. Companies are more likely to be subject to litigation, as a company is legally responsible for the actions of their employees in the workplace, and a company is a wealthier entity than one of its employees. The legal aspects of licensing software may protect software companies from litigation. However, greater legal protection for the customer can be built into the contract between the supplier and the customer for bespoke-software development.

[9]Many software companies have a defined code of ethics that employees are expected to adhere. Larger companies will wish to project a good corporate image and to be respected worldwide.

[10]The British Computer Society (BCS) has introduced a qualification system for computer science professionals that it used to show that professionals are properly qualified. The most important of these is the BCS Information Systems Examination Board (ISEB), which allows IT professionals to be qualified in service management, project management, software testing and so on.

[11]Software companies that are following the CMMI or ISO 9001 standards will employ audits to verify that the processes and procedures have been followed. Auditors report their findings to management and the findings are addressed appropriately by the project team and affected individuals.

9.3   Challenges in Software Engineering

The challenge in software engineering is to deliver high-quality software on time and on budget to customers. The research done by the Standish Group was discussed earlier in this chapter, and their 1998 research (Fig. 9.2) on project cost overruns in the US indicated that 33% of projects are between 21 and 50% overestimate, 18% are between 51 and 100% overestimate and 11% of projects are between 101 and 200% overestimate.

The accurate estimation of project cost, effort and schedule is a challenge in software engineering. Therefore, project managers need to determine how good their estimation process is and to make appropriate improvements. The use of software metrics is an objective way to do this, and improvements in estimation will be evident from a reduced variance between the estimated and actual effort. The project manager will determine and report the actual versus estimated effort and schedule for the project.

Risk management is an important part of project management, and the objective is to identify potential risks early and throughout the project, and to manage them appropriately. The probability of each risk occurring and its impact is determined and the risks are managed during project execution.

Software quality needs to be properly planned to enable the project to deliver a quality product. Flaws with poor quality software lead to a negative perception of the company, and may damage the customer relationship leading to a loss of market share.

There is a strong economic case for building quality into the software, as less time is spent in reworking defective software. The cost of poor quality (COPQ) should be measured and targets set for its reductions. It is important that lessons are learned during the project and are acted upon appropriately. This helps to promote a culture of continuous improvement.

There have been several high-profile software failures (O'Regan 2014) such as the millennium bug (Y2K) problem; the floating-point bug in the Intel

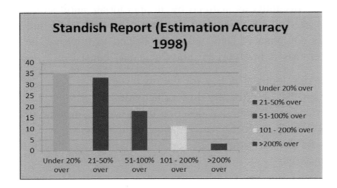

Fig. 9.2 Standish 1998 report—estimation accuracy

microprocessor in the mid-1990s; and the European Space Agency Ariane-5 disaster in 1996. The millennium bug was due to the use of two digits to represent dates rather than four digits. The solution involved finding and analysing all code that had a Y2K impact; planning and making the necessary changes; and verifying the correctness of the changes. The worldwide cost of correcting the millennium bug is estimated to have been in billions of dollars.

The Intel Corporation was slow to acknowledge the floating-point problem in its Pentium microprocessor, and in providing adequate information on its impact on its customers. This led to a large financial cost in replacing microprocessors for its customers. The Ariane-5 failure caused major embarrassment and damage to the credibility of the European Space Agency (ESA). Its maiden flight ended in failure on 4 June 1996, after a flight time of just 40 s.

These failures indicate that quality needs to be carefully considered when designing and developing software. The effect of software failure may be large costs to correct the software, loss of credibility of the company or even loss of life.

9.4 Software Processes and Life Cycles

Organisations vary by size and complexity, and the processes employed will reflect the nature of their business. The development of software involves many processes such as those for defining requirements, processes for project management and estimation, processes for design, implementation, testing and so on.

It is important that the processes employed are fit for purpose, and a key premise in the software quality field is that the quality of the resulting software is influenced by the quality and maturity of the underlying processes, and compliance to them. Therefore, it is necessary to focus on the quality of the processes, as well as the quality of the resulting software.

There is, of course, little point in having high-quality processes unless their use is institutionalised in the organisation. That is, all employees need to follow the processes consistently. This requires that people are trained on the new processes and that process discipline is instilled by an appropriate audit strategy.

Employees need to be trained on the processes, and audits are conducted to ensure process compliance. Data will be collected to improve the process. The software process assets in an organisation generally consist of:

- A software development policy for the organisation.
- Process maps that describe the flow of activities.
- Procedures and guidelines that describe the processes in more detail.
- Checklists to assist with the performance of the process.
- Templates for the performance of specific activities (e.g. design, testing).
- Training materials.

The processes employed to develop high-quality software generally include:

- Project management process,
- Requirements process,
- Design process,
- Coding process,
- Peer review process,
- Testing process,
- Supplier selection processes,
- Configuration management process,
- Audit process,
- Measurement process,
- Improvement process,
- Customer support and maintenance processes.

The software development process has an associated life cycle that consists of various phases. There are several well-known life cycles employed such as the waterfall model, the spiral model (Boehm 1988), the Rational Unified Process (Jacobson et al. 1999) and the Agile methodology which has become popular in recent years. The choice of the software development life cycle is determined from the needs of the specific project, and various life cycles are described in more detail in the following sections.

9.4.1   Waterfall Life Cycle

The origins of the waterfall model[12] (Fig. 9.3) are in the manufacturing and construction industry, and Winston Royce defined it formally for software development in 1970 (Royce 1970). It starts with requirements gathering and definition. It is followed by the functional specification, the design and implementation of the software and comprehensive testing. The testing generally includes unit, system and user acceptance testing.

It is employed for projects where the requirements can be identified early in the project life cycle or are known in advance. It is also called the 'V' life cycle model, with the left-hand side of the 'V' detailing requirements, specification, design and coding and the right-hand side detailing unit tests, integration tests, system tests and acceptance testing. Each phase has entry and exit criteria that must be satisfied before the next phase commences.

Many companies employ a set of templates to enable the activities in the various phases to be consistently performed. Templates may be employed for project planning and reporting requirements definition, design, testing and so on. These templates may be based on the IEEE standards or on industrial best practice.

---

[12]We treat the waterfall model as identical to the V model in this text.

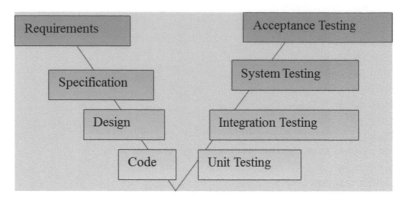

Fig. 9.3 Waterfall versus life cycle model

9.4.2 Spiral Life Cycles

The spiral model (Fig. 9.4) was developed by Barry Boehm in the mid-1980s, and it is useful when the requirements are not fully known at project initiation, or where the requirements evolve as a part of the development life cycle. The software development involves several spirals, where each spiral typically involves objectives and an analysis of the risks, updates to the requirements, design, code, testing and a user review of the iteration or spiral. The early spirals are concerned with prototyping, and the later spirals are concerned with the full implementation of the system.

The spiral is, in effect, a reusable prototype with the business analysts and the customer reviewing the current iteration, and providing feedback to the development team. The feedback is analysed and used to plan the next iteration. This approach is often used in joint application development, where the usability and look and feel of the application is a key concern. This is important in web-based development and in the development of a graphical user interface (GUI). The implementation of part of the system helps in gaining a better understanding of the requirements of the system, and this feeds into subsequent development cycles. The process repeats until the requirements and the software product are fully complete.

There are several variations of the spiral model including rapid application development (RAD), joint application development (JAD) models, and the dynamic systems development method (DSDM) model. Agile methods have become popular in recent years and these generally employ sprints (or iterations) of 2 weeks duration to implement several user stories.

There are other life cycle models, for example the iterative development process that combines the waterfall and spiral life cycle model. The cleanroom approach developed by Harlan Mills at IBM includes a phase for formal specification, and its approach to software testing is based on the predicted usage of the software product. The Rational Unified Process is discussed in the next section.

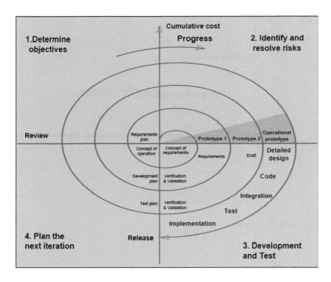

Fig. 9.4 SPIRAL life cycle model. Public domain

9.4.3 Rational Unified Process

The *rational unified process* (Jacobson et al. 1999) was developed at the Rational Corporation (now part of IBM) in the late 1990s. It uses the unified modelling language (UML) as a tool for specification and design, where UML is a visual modelling language for software systems that provide a means of specifying, constructing and documenting the object-oriented system. UML was developed by James Rumbaugh, Grady Booch and Ivar Jacobson, and it facilitates understanding of the architecture and complexity of the system.

RUP is *use case driven*, *architecture centric*, *iterative* and *incremental*, and includes cycles, phases, workflows, risk mitigation, quality control, project management and configuration control. Software projects may be very complex, and there are risks that requirements may be incomplete, or that the interpretation of a requirement may differ between the customer and the project team.

Requirements are gathered as use cases, which *describe the functional requirements from the point of view of the user of the system.* They describe what the system will do at a high level, and ensure that there is an appropriate focus on the user when defining the scope of the project. *Use cases also drive the development process*, as the developers create a series of design and implementation models that realise the use cases. The developers review each successive model for conformance to the use-case model, and the test team verifies that the implementation correctly implements the use cases.

The software architecture concept embodies the most significant static and dynamic aspects of the system. The architecture grows out of the use cases and

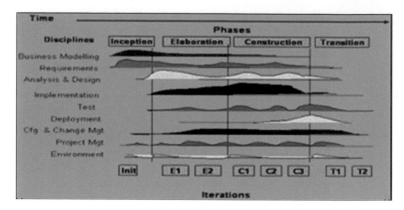

Fig. 9.5 Rational unified process

factors such as the platform that the software is to run on, deployment considerations, legacy systems and non-functional requirements.

RUP decomposes the work of a large project into smaller slices or mini-projects, and *each mini-project is an iteration that results in an increment to the product.* The iteration consists of one or more steps in the workflow, and generally leads to the growth of the product. If there is a need to repeat an iteration, then all that is lost is the misdirected effort of one iteration, rather than the entire product. That is, RUP is a way to mitigate risk in software engineering (Fig. 9.5).

9.4.4   Agile Development

There has been a growth of popularity among software developers in lightweight methodologies such as *Agile*. This is a software development methodology that claims to be more responsive to customer needs than traditional methods such as the waterfall model. *The waterfall development model is like a wide and slow-moving value stream*, and halfway through the project 100% if the requirements are typically 50% done. *However, for agile development 50% of requirements are typically 100% done halfway through the project.*

An early version of the methodology was introduced in the late 1980s/early 1990s, and the Agile Manifesto was introduced in early 2001 (www.agilealliance. org). Agile has a strong collaborative style of working and its approach includes:

- Aim is to achieve a narrow fast flowing value stream.
- Feedback and adaptation are employed in decision-making.
- User stories and sprints are employed.
- Stories are either done or not done.
- Iterative and incremental development is employed.
- A project is divided into iterations.

- An iteration has a fixed length (i.e. time boxing is employed).
- Entire software development life cycle is employed in the implementation of each story.
- Change is accepted as a normal part of life in the Agile world.
- Delivery is made as early as possible.
- Maintenance is considered part of the development process.
- Refactoring and evolutionary design employed.
- Continuous integration is employed.
- Short cycle times.
- Emphasis on quality.
- Stand-up meetings
- Plan regularly.
- Direct interaction preferred over documentation.
- Rapid conversion of requirements into working functionality.
- Demonstrate value early.
- Early decision-making.

Ongoing changes to requirements are considered normal in the Agile world, and it is believed to be more realistic to change requirements regularly during the project rather than attempting to define all the requirements at the start of the project. The methodology includes controls to manage changes to the requirements, and good communication and regular feedback is an essential part of the process.

A story may be a new feature or a modification to an existing feature. It is reduced to the minimum scope that can deliver business value, and a feature may give rise to several stories. Stories often build upon other stories and the entire software development life cycle is employed for the implementation of each story. *Stories are either done or not done*: i.e. *there is such thing as a story being 80% done.* The story is complete only when it passes its acceptance tests. Stories are prioritised based on several factors including:

- Business value of story
- Mitigation of risk
- Dependencies on other stories.

Sprint planning is performed before the start of the iteration, and stories are assigned to the iteration to fill the available time. The estimates for each story and their priority are determined, and the prioritised stories are assigned to the iteration. *A short morning stand-up meeting is held daily* during the iteration, and attended by the project manager and the project team. It discusses the progress made the previous day, problem reporting and tracking, and the work planned for the day ahead. A separate meeting is held for issues that require more detailed discussion.

Once the iteration is complete, the latest product increment is demonstrated to an audience including the product owner. This is to receive feedback and to identify new requirements. The team also conducts a retrospective meeting to identify what went well and what went poorly during the iteration, and this is used for continuous improvement in future iterations.

Agile employs pair programming and a collaborative style of working with the philosophy that two heads are better than one. This allows multiple perspectives in decision-making and a broader understanding of the issues.

Software testing is very important and Agile generally employs automated testing for unit, acceptance, performance and integration testing. Tests are run frequently with the goal of catching programming errors early. They are generally run on a separate build server to ensure that all dependencies are checked. Tests are rerun before making a release. *Agile employs test-driven development with tests written before the code.* The developers write code to make a test pass with ideally developers only coding against failing tests. This approach forces the developer to write testable code.

Refactoring is employed in Agile as a design and coding practice. The objective is to change how the software is written without changing what it does. Refactoring is a tool for evolutionary design where the design is regularly evaluated, and improvements are implemented as they are identified. The automated test suite is essential in showing that the integrity of the software is maintained following refactoring.

Continuous integration allows the system to be built with every change. Early and regular integration allows early feedback to be provided. It also allows all automated tests to be run thereby identifying problems earlier.

9.5 Activities in Waterfall Life Cycle

This section describes the various activities in the waterfall software development life cycle in more detail. The activities discussed include:

- Business Requirements Definition,
- Specification of System Requirements,
- Design,
- Implementation,
- Unit Testing,
- System Testing,
- UAT Testing,
- Support and Maintenance.

9.5.1 Business Requirements Definition

The business requirements specify what the customer wants, and define what the software system is required to do (*as distinct from how this is to be done*). The requirements are the foundation for the system, and if they are incorrect, then the implemented system will be incorrect. *Prototyping may be employed* to assist in the definition and validation of the requirements.

The specification of the requirements needs to be unambiguous to ensure that all parties involved in the development of the system share a common understanding of what is to be developed and tested.

Requirements gathering involve meetings with the stakeholders to gather all relevant information for the proposed product. The stakeholders are interviewed, and requirements workshops conducted to elicit the requirements from them. An early working system (prototype) is often used to identify gaps and misunderstandings between developers and users. The prototype may serve as a basis for writing the specification.

The requirements workshops are used to discuss and prioritise the requirements, as well as identifying and resolving any conflicting requirements. The collected information is consolidated into a coherent set of requirements.

The requirements are validated by the stakeholders to ensure that they are actually those desired, and to establish their feasibility. This may involve several reviews of the requirements until all stakeholders are ready to approve the requirements. Changes to the requirements may occur during the project, and these need to be controlled. It is essential to understand the impacts of a change prior to its approval.

The requirements for a system are generally documented in a natural language such as 'English'. Other notations that may be employed to express the requirements include the visual modelling language UML (Jacobson et al. 2005), and formal specification languages such as VDM or Z.

9.5.2 Specification of System Requirements

The specification of the system requirements of the product is essentially a statement of what the software development organisation will provide to meet the business requirements. That is, the detailed business requirements are a statement of what the customer wants, whereas the specification of the system requirements is a statement of what will be delivered by the software development organisation.

It is essential that the system requirements are valid with respect to the business requirements, and the stakeholders review them to ensure their validity. Traceability may be employed to show how the business requirements are addressed by the system requirements

There are two categories of system requirements: namely, functional and non-functional requirements. The *functional requirements* define the functionality that is required of the system, and it may include screenshots, report layouts or the desired functionality specified in natural language, use cases, etc. The *non-*

functional requirements will generally include security, reliability, performance and portability requirements, as well as usability and maintainability requirements.

9.5.3   Design

The design of the system consists of engineering activities to describe the architecture or structure of the system, as well as activities to describe the algorithms and functions required to implement the system requirements. It is a creative process concerned with how the system will be implemented, and its activities include architecture design, interface design and data structure design. There are often several possible design solutions for the proposed system, and the designer will need to decide on the most appropriate solution.

The design may be specified in various ways such as graphical notations that display the relationships between the components making up the design. The notation may include flowcharts, or various UML diagrams such as sequence diagrams, state charts and so on. Program description languages or pseudocode may be employed to define the algorithms and data structures that are the basis for implementation.

Functional design involves starting with a high-level view of the system and refining it into a more detailed design. The system state is centralised and shared between the functions operating on that state.

Object-oriented design is based on the concept of *information hiding* (Parnas 1972). The system is viewed as a collection of objects rather than functions, with each object managing its own state information. The system state is decentralised and an object is a member of a class. The definition of a class includes attributes and operations on class members, and these may be inherited from superclasses. Objects communicate by exchanging messages.

It is essential to verify and validate the design with respect to the system requirements, and this will be done by design reviews, and traceability of the design to the system requirements.

9.5.4   Implementation

This phase is concerned with implementing the design in the target language and environment (e.g. C++ or Java), and involves writing or generating the actual code. The development team divides up the work to be done, with each programmer responsible for one or more modules. The coding activities include code reviews or walkthroughs to ensure that quality code is produced, and to verify its correctness. The code reviews will verify that the source code adheres to the coding standards, that maintainability issues are addressed, and that the code produced is a valid implementation of the software design.

Software reuse has become more important in recent times as it provides a way to speed up the development process. Components or objects that may be reused need to be identified and handled accordingly. The implemented code may use software components that have either been developed internally or purchased off the shelf. Open-source software has become popular in recent years, and it allows software developed by others to be used (*under an open-source license*) in the development of applications.

The benefits of software reuse include increased productivity and a faster time to market. There are inherent risks with customized-off-the shelf (COTS) software, as the supplier may decide to no longer support the software, or there is no guarantee that software that has worked successfully in one domain will work correctly in a different domain. It is, therefore, important to consider the risks as well as the benefits of software reuse and open-source software.

9.5.5   Software Testing

Software testing is employed to verify that the requirements have been correctly implemented, and that the software is fit for purpose, as well as identifying defects present in the software. There are various types of testing that may be conducted including *unit testing, integration testing, system testing, performance testing and user acceptance testing*. These are described below:

Unit Testing

Unit testing is performed by the programmer on the completed unit (or module), and prior to its integration with other modules. The programmer writes these tests, and the objective is to show that the code satisfies its design. Each unit test case is documented and it should include a test objective and the expected result.

Code coverage and branch coverage metrics are often recorded to give an indication of how comprehensive the unit testing has been. These metrics provide visibility into the number of lines of code executed as well as the branches covered during unit testing.

The developer executes the unit tests, records the results, corrects any identified defects and retests the software. *Test-driven development* has become popular in recent years (e.g. in the Agile world), and this involves writing the unit test case before the code, and the code is written to pass the unit test cases.

Integration Test

The development team performs this type of testing on the integrated system, once all individual units work correctly in isolation. The objective is to verify that the modules and their interfaces work correctly together, and to identify and resolve any issues. Modules that work correctly in isolation may fail when integrated with other modules.

System Test

The purpose of system testing is to verify that the implementation is valid with respect to the system requirements. It involves the specification and execution of system test cases to verify that the system requirements have been correctly implemented. An independent test group generally conducts this type of testing, and the system tests are traceable to the system requirements.

Any system requirements that have been incorrectly implemented will be identified, and defects logged and reported to the developers. The test group will verify that the revised version of the software is correct, and regression testing is carried out to verify system integrity. System testing may include security testing, usability testing and performance testing.

The preparation of the test environment may involve ordering special hardware and tools, and it is important that the test environment is set up as early as possible to allow the timely execution of the test cases.

Performance Test

The purpose of performance testing is to ensure that the performance of the system is within the bounds specified in the non-functional requirements, and to determine if the system is scalable to support future growth. It may include *load performance testing*, where the system is subjected to heavy loads over a long period, and *stress testing*, where the system is subjected to heavy loads during a short time interval.

Performance testing often involves the simulation of many users using the system, and measuring the response times for various activities. Test tools are employed to simulate many users and heavy loads.

User Acceptance Test

UAT testing is usually performed under controlled conditions at the customer site, and its operation will closely resemble the real-life behaviour of the system. The customer will see the product in operation, and can judge if the system is fit for purpose.

The objective is to demonstrate that the product satisfies the business requirements and meets the customer expectations. Upon its successful completion, the customer should be happy to accept the product.

9.5.6   Maintenance

This phase continues after the release of the software product to the customer. Any problems that the customer notes with the software are reported as per the customer support and maintenance agreement. The support issues will require investigation, and the issue may be *a defect in the software, an enhancement to the software* or *due to a misunderstanding*. The support and maintenance team will identify the causes of any identified defects, and will implement an appropriate solution. Testing is conducted to verify that the solution is correct, and that the changes made have not adversely affected other parts of the system. Mature organisations will conduct postmortems to learn lessons from the defect[13], and will take corrective action to prevent a reoccurrence.

The presence of a maintenance phase suggests an acceptance of the reality that problems with the software will be identified post release. The goal of developing a correct and reliable software product the first time is very difficult to achieve, and the customer is always likely to find some issues with the released software product. It is accepted today that quality needs to be built into each step in the development process, with the role of software inspections and testing to identify as many defects as possible prior to release, and minimise the risk that that serious defects will be found post release.

The more effective the in-phase inspections of deliverables, the higher the quality of the resulting implementation, with a corresponding reduction in the number of defects detected by the test groups. The testing group plays a key role in verifying that the system is correct, and in providing confidence that the software is fit for purpose. Testing and retesting to achieve quality, until the testing group is confident that all defects have been eliminated almost seems to be a '*brute force*' approach. Dijkstra (1972) noted that:

Testing a program demonstrates that it contains errors, never that it is correct.

That is, irrespective of the amount of time spent testing, it can never be said with absolute confidence that the program is correct, and, at best, statistical techniques may be employed to give a measure of the confidence in its correctness. That is, there is no guarantee that all defects have been found in the software.

Many software companies may consider one defect per thousand lines of code (KLOC) to be reasonable quality. However, if the system contains one million lines of code this is equivalent to a thousand post-release defects, which is unacceptable.

Some mature organisations have a quality objective of three defects per million lines of code. This goal is known as Six Sigma (6σ), and Motorola developed it initially for its manufacturing businesses and later applied to its software organisation. The goal is to reduce variability in manufacturing processes and to ensure that the processes performed within strict process control limits. Motorola was awarded the first Malcolm Baldridge Quality Award for its Six Sigma initiative and its commitment to quality.

---

[13]This is essential for serious defects that have caused significant inconvenience to customers (e.g. a major telecom outage). The software development organisation will wish to learn lessons to determine what went wrong in its processes that prevented the defect from been identified during peer reviews and testing. Actions to prevent a reoccurrence will be identified and implemented.

9.6   Software Inspections

Software inspections are used to build quality into software products, and there are several well-known approaches such as the Fagan Methodology (Fagan 1976) developed by Michael Fagan of IBM. This is a seven-step process that identifies and removes errors in work products. The process mandates that requirement documents, design documents, source code and test plans are all formally inspected by experts who are independent of the author of the deliverable.

There are various *roles* defined in the process including the *moderator* who chairs the inspection. The *reader's* responsibility is to read or paraphrase the deliverable, and *the author* is the creator of the deliverable and the *tester* role is concerned with the testing viewpoint.

The inspection process will consider whether the design is correct with respect to the requirements, and whether the source code is correct with respect to the design. Software inspections play an important role in building quality into the software, and in reducing the cost of poor quality in the organisation. For more detailed information on software inspections see (O'Regan 2014).

9.7   Software Project Management

The timely delivery of quality software requires good project management and engineering processes. Software projects have a history of being delivered late or overbudget, and good project management practices include activities such as:

- Initiating and planning the project.
- Estimation of cost, effort and schedule for the project.
- Preparing the initial project plan and schedule.
- Definition of the key milestones.
- Obtaining approval for the project plan and schedule from the stakeholders.
- Identifying and managing risks.
- Staffing the project.
- Managing project execution.
- Monitoring and managing progress, budget, schedule, effort, risks, issues, change requests and quality.
- Taking corrective action.
- Replanning and rescheduling the project.
- Communicating progress to the stakeholders.
- Preparing status reports and presentations.
- Closing the project.

The project plan will contain or reference several other plans such as the project quality plan, the communication plan, the configuration management plan and the test plan.

Project estimation and scheduling are difficult for software projects as often these involve new technologies and are breaking new ground. This means that they are often may be quite different from previous projects, and so historical estimates may not be a good basis for estimation. Further, unanticipated problems may arise with technically advanced projects, and so the estimates may be overly optimistic.

Gantt charts are generally employed for project scheduling, and these show the work breakdown for the project, as well as task dependencies. The Gantt chart shows the allocation of staff to the various tasks, where each task has a start date and an end date, the effort associated with it, as well as the staff involved.

The effective management of risk during a project is essential to project success. Risks arise due to uncertainty and the risk management cycle involves[14] risk identification; risk analysis and evaluation, identifying responses to risks, selecting and planning a response to the risk and risk monitoring. The risks are logged, and the likelihood of each risk arising and its impact is then determined. The risk is assigned an owner and an appropriate response to the risk determined. For more detailed information on project management see (O'Regan 2017a).

9.8   CMMI Maturity Model

The CMMI is a framework to assist an organisation in the implementation of best practice in software and systems engineering (Chrissis et al. 2011). It is an internationally recognised model for process improvement and assessment, and is used worldwide by thousands of organisations. It provides a framework for an organisation to introduce a solid engineering approach to the development of software, and the CMMI practices support the implementation of high-quality processes for the various software engineering and management activities.

It was developed by the Software Engineering Institute (SEI), who adapted the process improvement principles used in the manufacturing field to the software field. It developed the original CMM model in the early 1990s, and its successor is the CMMI. The CMMI states *what the organisation needs to do* to mature its processes rather than *how this should be done*, and so the organisation has the freedom to interpret the model meets its business needs effectively.

The CMMI consists of five maturity levels with each maturity level consisting of several process areas. Each process area consists of a set of goals, which are implemented by practices related to that process area. Level two is focused on management practices; level three is focused on engineering and organisation practices; level four is concerned with ensuring that key processes are performing within strict quantitative limits; level five is concerned with continuous process improvement. Maturity levels may not be skipped in the staged implementation of the CMMI, as each maturity level provides the foundation for the next level.

---

[14]These are the risk management activities in the Prince2 methodology.

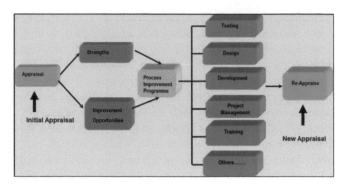

Fig. 9.6 Software process improvement

The CMMI allows organisations to benchmark themselves against other organisations. This is done by a formal appraisal conducted by an authorised lead appraiser (Standard CMMI Appraisal Method for Process Improvement 2006). The results of the appraisal are generally reported back to the SEI, and there is a strict qualification process to become an *authorised lead appraiser.*

An appraisal is useful in that it allows the organisation to determine its current software process maturity. It may be used to verify that the organisation has improved, and it enables the organisation to prioritise improvements (Fig. 9.6). The CMMI is discussed in more detail in (O'Regan 2010, 2014).

9.9   Formal Methods

Dijkstra and Hoare argued that the appropriate way to develop correct software is to derive the program from its formal mathematical specification, and to employ *mathematical proof* to demonstrate the correctness of the software with respect to the specification. This is a rigorous framework to develop programs adhering to the highest quality constraints. However, in practice mathematical techniques have proved to be cumbersome to use, and their widespread deployment in industry is unlikely at this time.

The *safety-critical area* is one domain to which mathematical techniques have been successfully used. For example, it is essential in the railway domain that a property such as *'when a train is in a level crossing, then the gate is closed'* is demonstrated to be correct, and formal methods can play a key role in the verification of safety-critical properties. There is a need for extra rigour in the software development process in the safety-critical field, and mathematical techniques can demonstrate the presence or absence of certain desirable or undesirable properties.

Spivey (1992) defines a *'formal specification'* as the use of mathematical notation to describe in a precise way the properties which an information system must have, without unduly constraining the way in which these properties are achieved. It describes *what* the system must do, as distinct from *how* it is to be done. This abstraction away from implementation enables questions about what the system does to be answered, independently of the detailed code. Further, the unambiguous nature of mathematical notation avoids the problem of speculation about the meaning of phrases in an imprecisely worded natural language description of a system.

The formal specification thus becomes the key reference point for the different parties concerned with the construction of the system, and is a useful way of promoting a common understanding for all those concerned with the system.

The term *'formal methods'* is used to describe a formal specification language and a method for the design and implementation of computer systems. The specification is written in a mathematical language, and its precision helps to avoid the problem of ambiguity inherent in a natural language specification.

The derivation of the software from the specification may be achieved via *stepwise refinement*. Each refinement step makes the specification more concrete and closer to the actual implementation. There is an associated *proof obligation* that the refinement be valid, and that the concrete state preserves the properties of the abstract state. Thus, assuming the original specification is correct and the proofs of correctness of each refinement step are valid, then there is a very high degree of confidence in the correctness of the implemented software.

Formal methods have been applied to a diverse range of applications, including circuit design, artificial intelligence, specification of standards, specification and verification of programs, etc. They are described in more detail in (O'Regan 2017b).

9.10 Review Questions

1. Discuss the research results of the Standish Group on the current state of IT project delivery.
2. What are the main challenges in software engineering?
3. Describe the various existing software life cycles
4. What are the advantages and disadvantages of Agile?
5. Describe the purpose of software inspections? What are the benefits?
6. Describe the main activities in software testing.
7. Describe the advantages and disadvantages of formal methods.
8. Describe the main activities in project management.
9. Explain the significance of the CMMI.

9.11 Summary

The birth of software engineering was at the NATO conference held in 1968 in Germany. This conference highlighted the problems that existed in the software sector in the late 1960s, and the term '*software crisis*' was coined to refer to these. It led to the realisation that software engineers need to be properly trained to enable them to build high-quality products that are safe to use.

The Standish Group conducts research on the extent of problems with the delivery of projects on time and budget. Their research indicates that it remains a challenge to deliver projects on time, on budget and with the right quality.

Programmers are like engineers in the sense that they build products. Therefore, programmers need to receive an appropriate education in engineering as well as on the latest technologies. Classical engineers receive training on product design, and an appropriate level of mathematics.

Software engineering involves multi-person construction of multi-version programs. It is a systematic approach to the development and maintenance of the software, and it requires a precise statement of the requirements of the software product, and then the design and development of a solution to meet these requirements. The solution is verified by rigorous software testing.

Software process maturity models such as the CMMI place an emphasis on understanding and improving the software processes in an organisation. The CMMI is a framework to implement high-quality processes, and a SCAMPI appraisal allows organisations to benchmark themselves against other similar organisations.

Formal methods involve the use of mathematical techniques to provide extra confidence in the correctness of the software. They are used mainly in the safety- and security-critical field.

Chapter 10
Overview of Operating Systems

10.1 Introduction

An *operating system* is a collection of software programs that control the hardware of a computer and makes it usable. It makes the computing power of the hardware available to the users of the computer, and it manages the hardware to achieve good system performance. An operating system manages system hardware such as the processors, storage, input/output devices, communication devices and data, and it provides functionality such as sharing hardware among users, scheduling resources among users, preventing users from interfering with each other, facilitating input/output, recovering from errors and handling network communication.

The earliest computers did not have an operating system, and the user had exclusive control over a large computer for a specified period. The user entered the program one bit at a time in machine code (initially using mechanical switches and later with a stack of punched cards), and waited for the results. People began to develop libraries to share the code for common activities, and these are in a sense the precursor of today's operating systems.

The earliest operating systems were designed in the 1950s with the goal of making more efficient use of expensive computer resources. These batch-processing

© Springer International Publishing AG, part of Springer Nature 2018 203
G. O'Regan, *World of Computing*,
https://doi.org/10.1007/978-3-319-75844-2_10

systems ran one job at a time, and programs and data were submitted in groups (or batches).

These evolved during the early 1960s into multi-batch systems that were designed to improve utilisation of the expensive computer resources. They could handle several diverse jobs at once, which offered a way to optimise computer utilisation. One job could be using the processor while another job could be using the various I/O devices. These later batch-processing systems contained many peripheral devices such as card readers, card punches, printers, tape drives and disk drives. Jobs were normally submitted on punched cards and computer tape, and often a user's job could sit for hours (days) on an input table until it was processed. However, even a very slight error in a program would cause the program to fail, and it would require resubmission. This meant that software development in this environment was very slow.

This led operating system designers to develop the concept of *multiprogramming* in which several jobs are in main memory at once, and the concept of interrupts, where an *interrupt* allows one unit to gain the attention of another. The state of the interrupted unit is saved prior to the processing of the interrupt, and restored once processing is complete.

MIT developed the CTSS time-sharing system in the early 1960s, and this operating system provided users with typewriter-like terminals to obtain computing power from the machine. CTSS ran a conventional batch stream (to ensure high utilisation of expensive computer resources), but it was also able to give fast responses to users who were editing or debugging programs. It was a highly interactive environment where the computer provided rapid responses to user requests.

IBM announced the System/360 family of computers in 1964, and the computers in the family were designed to use the IBM System/360 operating system (OS/360). OS/360 was a batch-oriented operating system, and IBM supported three variants of OS/360, which allowed multiprogramming for the mid-range and top-range members of the family.

The other major operating system used in the System 360 was the disk operating system (DOS/360)[1]. The IBM System/360 evolved over time into the System/370 series. IBM began work on the CP/CMS operating system in 1964, and this time-sharing operating system was popular in the 1960s. It evolved into IBM's VM operating system in the early 1970s.

MIT's successor to the CTSS operating system was a general time-sharing operating system called *"Multics"*, and Bell Labs was initially involved in its development. UNIX arose out of work on the development of Multics, and it was developed at Bell Labs in the early 1970s. It is a multitasking and multi-user operating system.

The IBM PC was introduced in 1981, and IBM outsourced the development of the operating system to a small company called Microsoft. The terms of the deal with IBM allowed Microsoft the right to license its version of the operating system,

---

[1]Not to be confused with DOS which was used on the original IBM personal computers.

MS/DOS, on IBM compatibles, with PC/DOS (or simply DOS) reserved for IBM personal computers only. MS/DOS managed floppy disks and files, input and output, memory and it contained an external command processor that interpreted user commands, and allowed the user to interact with the system.

The Macintosh was a paradigm shift in the computer industry when it was introduced in 1984. Its MAC operating system was GUI based, friendly, intuitive and easy to use, and it was clear that the future of operating systems was in GUI-driven systems, rather than primitive command-driven operating systems such as MS/DOS.

Microsoft Windows is a family of graphical operating systems developed by Microsoft, and it was Microsoft's initial response to Apple's GUI operating system. Windows has evolved to become the dominant operating system on laptops and personal computers, but it has failed to make an impact on the smartphone operating system market, which is dominated by Apple's iOS and Google's Android operating systems.

The Android operating system was designed mainly for touchscreen smartphones and tablets, and it was developed by Google and the Open Handset Alliance. Android is built on the Linux kernel, and its first version was released in late 2007.

The iOS operating system is a mobile operating system employed on Apple's mobile devices such as smartphones and tablets. It was introduced in 2007. For more detailed information on operating systems see (Deitcl 1990; Anderson and Dahlin 2014).

10.2 Fundamentals of Operating Systems

An operating system is a collection of software programs that control the hardware of a computer and makes it usable. The operating system may be dealing with a single processor or multiprocessor system. The concept of a *process* (program in execution) is central to understanding modern operating systems, and a process goes through a series of discrete process states with an *event* leading to a change of state. A process is said to be *running* if it currently has the CPU, and a process is said to be *ready* if it could use a CPU (should one be available). A process is said to be *blocked* if it is waiting for some event to happen (e.g. an *i/o* event) before it can continue (Fig. 10.1).

A process is created in response to the submission of a job to the system, and it is generally added to the end of the ready list. The process then gradually moves to the head of the ready queue, and when the CPU becomes available the process makes a transition from the ready state to the running space. The assignment of the CPU to the first process on the ready list is termed *dispatching*, and the operating system sets a hardware interrupting clock to allow the process to run for a fixed period (or *quantum*), and the operating system interrupts (where appropriate) to ensure that the next ready process is dispatched to running before (or at) the end of the time quantum.

Fig. 10.1 Process state transitions

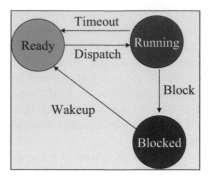

The *process control block* (PCB) is a data structure containing key information about the process including its current state, priority and pointers to parent and child processes (i.e. the process that created it and any processes that it created). It defines the process to the operating system. Processes may be created or destroyed, suspended or resumed, blocked or woken up and dispatched. A suspended process cannot continue until another process resumes it, and the suspension may be initiated by the process itself or another process.

An *interrupt* is an event that alters the sequence in which a processor executes instructions, and it is generated by the hardware of the computer system. It results in the operating system gaining control, and the state of the interrupted process is saved. The operating system then analyses the interrupt and passes control to the *interrupt handler* for processing the interrupt. Finally, the interrupted process is resumed.

Concurrency is a form of computing in which multiple computations (processes) are executed during the same time period. *Parallel computing* allows execution to occur in the same time instant (on separate processors of a multiprocessor machine), whereas concurrent computing consists of process lifetimes overlapping and where execution need not happen at the same time instant.

Concurrency employs *interleaving* where the execution steps of each process employ time-sharing slices so that only one process runs at a time, and if it does not complete within its time slice it is paused, another process begins or resumes, and then later the original process is resumed. That is, only one process is running at a given time instant, whereas multiple processes are part of the way through execution.

It is important to identify and deal with concurrency-specific errors such as deadlock and livelock. A *deadlock* is a situation in which the system has reached a state in which no further progress can be made, and at least one process needs to complete its tasks. Figure 10.2 illustrates a deadlock situation, where both processes are waiting for the other to free a resource that it will not free until the other frees its resource (circular wait). *Livelock* refers to a situation where the processes in a system are stuck in a repetitive task, and are making no progress towards their functional goals.

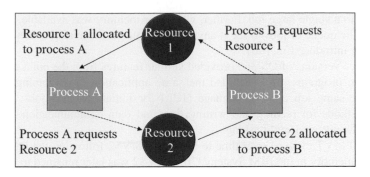

Fig. 10.2 A simple deadlock

It is essential that properties such as *mutual exclusion* (at most one process is in its critical section at any given time) should not be violated. The critical section refers to shared modifiable data, and so it must be ensured that when one process is in a critical section then all other processes that access the same shared modifiable data are excluded from their critical sections. One common implementation of mutual exclusion is by *semaphores* (a protected variable whose values may be accessed and modified only by the **P** and **V** operations).

It is essential that something bad (e.g. a deadlock situation) never happens, and that *liveness properties* (a desired event or something good eventually happens). It is essential that *invariants* (properties that are true all the time) are not violated.

10.3   IBM OS/360 and MVS

IBM announced the System/360 family of computers in 1964 (Fig. 3.20), and the family was designed to use the IBM System/360 operating system (OS/360). OS/360 was a batch-oriented operating system, and IBM supported three variants of it. These were OS/360 PCP (Principal Control Program), OS/360 MFT (Multiple Programming with a Fixed number of Tasks) and OS/360 MVT (Multiple Programming with a Variable number of Tasks).

OS/360 PCP was the simplest version, and it could run only one program at a time. The smaller members of the System/360 family used it. OS/360 MFT could run several programs at once, but only after partitioning the memory required to run each. It was subject to the limitation that if a program was idle, its allocated memory was unavailable to other programs. It was developed as an interim solution pending the delayed introduction of OS/360 MVT. However, the simpler MFT continued in use for many years due to problems with MVT.

OS/360 MVT was the most sophisticated version of OS/360, and it was intended for the largest members in the System/360 family. It allowed memory divisions to be re-created as needed, and it could allocate all the computer's memory (if required) to a single large job. Further, whenever memory was available, OS/MVT searched the queue of jobs to see if any could be run on the available memory. OS/MVT was introduced in 1967.

All three versions of OS/360 provided similar features from the point of view of application programs. This included the same application programming interface (API); the same job control language (JCL) for initiating batch jobs; the same access methods for reading and writing files and data communication; the same spooling facility and multitasking.

OS/360 MVT evolved over time to become OS/VS2 following the introduction of virtual memory in the IBM System/370. OS/VS2 was later renamed to OS/MVS.

IBM introduced the multiple virtual storage (MVS) operating system in 1974, and it was an enhancement of the MVT version of the OS/360 operating system that supported virtual memory. It was the most commonly used operating system on the IBM System/370 and System/390 mainframe computers.

The System/370 was an enhancement of the System/360 architecture in that it provided virtual storage capabilities, where *virtual storage* (memory) allows a much larger memory space to be addressed than is available in the primary memory of the computer. The concept of virtual storage dates to the design of the Atlas Computer at the University of Manchester in 1960, and the two most common methods of implementing virtual storage are paging and segmentation.

The 24-bit addressing of the System/370 meant that each user (or job) had a 16-megabyte (2^{24}) virtual address space (i.e. 256 segments, with each segment containing 16 pages, and each page contained 4096 bytes). The page table performs the translation between the virtual address as seen by the application into the physical address used by the hardware, and this may involve swapping pages from physical storage to main memory.

MVS provided multiprogramming and multiprocessing capabilities, and it was a large operating system designed with performance, reliability and availability in mind. The operating system had recovery routines that gained control in the event of an operating system failure, and it attempted recovery from hardware errors.

MVS included a master scheduler that initialised the system and responded to commands issued by the system operator. It contained a job entry subsystem for jobs to be entered, and its system management facility collected information to analyse system performance. Its time sharing option (TSO) provided users with interactive editing, testing and debugging capabilities, and its data management functionality handled all input/output and file management activities. Its telecommunication functionality allowed remote terminal users to access MVS.

10.4   VM

The virtual machine (VM) operating system makes a single machine appear as several real machines (Fig. 10.3). The user at a VM virtual machine has what seems to be a complete real machine, even though this is just an illusion. A virtual

Fig. 10.3 Virtual machine
operating system

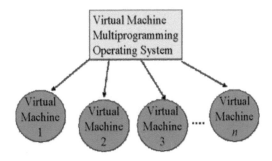

machine runs programs in a similar way to a real machine, and the user commu-
nicates with the virtual machine through a terminal. The most widely used virtual
machine operating system is IBM's VM, which was released in the early 1970s. It
was used on an IBM System/370 mainframe, and created the illusion that each user
operating at a terminal had access to a complete IBM 370, including its input/output
devices.

VM can run several different operating systems at once, each of them on its own
virtual machine. This is a very attractive feature as running multiple operating
systems offers a form of backup in the event of failure. The operating systems
running on virtual machines perform their normal functions such as storage management, control of input/output, processor scheduling and multiprogramming.

Virtual machines create virtual processors, virtual storage and virtual I/O devices. The VM user may run operating systems such as MVS, VM/370, AIX/370 or
VM itself.

The main components of VM are the control program (CP), the conversational
monitor system (CMS), the remote spooling communications subsystem (RSCS),
the interactive problem control system (IPCS) and the CMS Batch.

CP creates the environment in which virtual machines may execute, and it
provides support for the various operating systems that may be used to control the
IBM/370. It manages the real machine underlying the virtual machine environment,
and gives each user access to the facilities of the real machine.

CMS is an applications system with editors, debugging tools and various
application packages. RSCS provides the ability to transmit and receive files, and
IPCS is used for online analysis and for fixing VM software problems. The CMS
batch facility allows the user to submit longer jobs for batch processing.

10.5   VMS

The VAX virtual memory system (VMS) was designed as the operating system for
the VAX family of minicomputers. Digital Equipment Corporation (DEC) introduced the VAX family in the late 1970s, and DEC was a major player in the
minicomputer market with its popular PDP and VAX minicomputers. The models

in the VAX family of computers all had the same architecture, and they could all run the VMS operating system.

David Cutler and others at DEC designed VMS as a high-end, secure, scalable, multi-user, multitasking and virtual memory operating system, that supported a broad class of applications and systems. DEC developed VAX and VMS together, and the designers balanced the trade-offs between the work done by the hardware and the work done by the operating system.

VAXes may operate together in a peer-to-peer relationship, where any VAX may be a client or any may be a server. This allows flexibility when several computers perform tasks in cooperation. Several VAXes may be connected so that they work as a cooperating unit called a VAXcluster.

VMS expanded the memory of the machine by disk or other peripheral storage to act as extra memory. The VAX-11 provided a 32-bit virtual address space per process, divided into 512 byte pages. VMS used paging and segmentation, with the first 23 bits used as the virtual page number (VPN), and a 9-bit offset within the page.

VMS was a popular and easy-to-use operating system. Its commands are easy to remember English like words, and it has an extensive online help system. It included utilities such as a mail program and a text editor. Open VMS is the latest version of the operating system, and is sold by HP.

10.6   UNIX

Ken Thompson, Dennis Ritchie and others designed and developed the UNIX operating system at Bell Labs in the early 1970s. It is a multitasking and multi-user operating system that was written almost entirely in the C programming language (which was designed by Denis Ritchie at Bell Labs). UNIX arose out of work by Massachusetts Institute of Technology, General Electric and Bell Labs on the development of a general time-sharing operating system called "*Multics*".

Bell Labs decided in 1969 to withdraw from the Multics project (as they believed that it would be a large and expensive system), and to use General Electric's GECOS operating system. However, several of the Bell Lab researchers (led by Ken Thompson) decided to continue the work on a smaller scale operating system, and the name "UNIX" was coined by Brian Kernighan. The first version of the operating system was written on a Digital PDP-7 minicomputer in assembly language, and Dennis Ritchie joined the project. He helped in rewriting UNIX in the C programming language for the PDP-11 computer in 1973, which had recently been introduced. Thompson and Ritchie later received the Turing Award for their design and development of the UNIX operating system. Microsoft introduced XENIX, a commercial version of UNIX, in 1980.

The use of C helped to make UNIX more portable, and it became a widely used operating system. It was initially used by universities and the US government, and it later became popular in industry. It is a powerful and flexible operating system that is used on a variety of machines from micros to supercomputers. It is designed to allow several programmers to access the computer at the same time, and to share its resources, and it offers powerful real-time sharing of resources.

It includes features such as *multitasking* which allows the computer to do several things at once; *multi-user* capability which allows several users to use the computer at the same time; and *portability* of the operating system which allows it to be used on several computer platforms with minimal changes to the code. It includes a collection of tools and applications. There are three levels of the UNIX system namely *kernel*, *shell* and *tools and applications*.

The kernel is the central part of the UNIX operating system, and it provides systems services to applications programs. This includes services for process management, memory management, and input/output management. UNIX manages many concurrent processes.

The UNIX shell is a command interpreter that acts as the interface between the user and the operating system. There are several popular shells for UNIX including the Bourne shell and Korn shell. UNIX uses a hierarchical file system with the root node at its origin, with each directory entry containing files and other directories. For a more detailed account of UNIX see (Robbins 2005).

10.7 MS/DOS

We discussed the introduction of the IBM personal computer in Chap. 3, as well as the controversy with respect to the development of the PC/DOS operating system. Digital Research, the developers of the CP/M operating system, lost out on the major opportunity of supplying the operating system for the IBM PC, and instead it was Microsoft that reaped the benefits. The terms of the deal with IBM allowed Microsoft the right to license its operating system, MS/DOS, on IBM compatibles, whereas PC/DOS (or simply DOS) was reserved for IBM personal computers only.

The open architecture of the IBM PC led to the development of cheaper IBM compatible personal computers (clones of the IBM PC but cheaper), and they rapidly gained market share, as it was difficult for IBM to compete on price. This led to a massive international demand for MS/DOS (which was the operating system for IBM compatibles and clones).

The IBM PC was introduced in 1981, and the first version of the operating system was compatible with Digital Research's CP/M operating system (as it essentially was CP/M). It managed floppy disks and files, input and output, memory and it contained an external command processor that interpreted user commands, and allowed the user to interact with the system.

MS/DOS version 2.0 was introduced in 1983, and it was designed to support the 10 MB hard disk on the IBM PC/XT, as well as providing support for device drivers. Microsoft had previously licensed XENIX (their commercial version of UNIX) from AT&T, and MS/DOS 2.0 was a move towards XENIX. It employed a hierarchical file system, and a unique path name identified each file (like XENIX).

It provided limited multitasking for background print spooling. The hard disk on the XT helped to establish the IBM PC in the business marketplace.

MS/DOS 3.0 was released in 1984, and it provided support for the IBM PC/AT, which had a 20 MB hard disk. Several versions of MS/DOS followed through the 1980s and 1990s, and were used with Microsoft Windows 95 and Windows Millennium. Today, Microsoft Windows is the operating system used on personal computers, and development of MS/DOS ceased in 2000. It is now of historical interest.

10.8 Microsoft Windows

The Apple Macintosh was released in 1984, and its MAC operating system was GUI based and was a paradigm shift in the computer industry. It was friendly, intuitive and easy to use, and it was clear that the future of operating systems was in GUI-driven systems, rather than in the primitive command driven operating systems such as MS/DOS.

Microsoft Windows is a family of graphical operating systems developed by Microsoft. The original Windows 1.0 operating environment was introduced in late 1985 as a graphical operating system shell for its command-driven MS/DOS operating system. It was Microsoft's initial response to Apple's GUI operating system.

The early versions of Windows were not complete operating systems as such and were instead graphical shells in that they ran on top of MS/DOS and extended the operating system. Windows 1.0 used MS/DOS for file system services, and it also included applications such as a calculator, calendar and clock. However, Windows differed from MS/DOS in that it allowed multiple graphical applications to be run at the same time, and this was done through cooperative multitasking.

Windows 2.0 was introduced in 1987 and it was more popular than its predecessor. It included improvements to the user interface and to memory management. Windows 3.0 improved the design of the operating system, and it used virtual memory and virtual device drivers that allowed arbitrary devices to be shared between multitasked DOS applications. It was introduced in 1990 and it was the first Windows operating system to achieve commercial success.

Windows 3.1 was introduced in 1992, Windows 95 in 1995, Windows 98 in 1998 and Windows Millennium (ME) in 2000. Windows ME provided expanded multimedia capabilities including the Windows Media Player, and it was the last DOS-based version of Windows. Windows ME was criticised for its speed and instability.

Windows XP was introduced in 2001 and it was marketed as a "Home" edition for personal users and a "Professional" edition for business users. Windows Vista was released in 2006, Windows 7 in 2009, Windows 8 in 2012 and Windows 10 was released in 2015.

Microsoft Windows dominates the personal computer and laptop market with over 90% market share. Windows has not been as successful on mobile computing platforms such as mobile phones and tablets, where Google's Android operating system is the dominant platform.

10.9 Mobile Operating Systems

Android (Fig. 10.4) is a mobile operating system that was developed by Google and the Open Handset Alliance for touchscreen smartphones and tablets. It is built on the Linux kernel, and the first version of the operating system was released in late 2007. The first Android smartphone was released in late 2008, and Android is currently the most widely used operating system.

The source code for Android is released under an open-source license, and its open-source philosophy has led to a large community of developers who maintain and develop new versions of it. Manufacturers may modify Android as they see fit, and this allows them to customise their devices and differentiate them from competitor products.

There are over a million applications (apps) for Android, and developers are challenged to ensure that the apps are compatible with the many mobile devices using different hardware and running various (possibly customised) versions of Android.

The iOS operating system is a mobile operating system employed on Apple's mobile devices such as smartphones and tablets. It was created from the MAC OS/X operating system, and introduced in 2007. Multitasking for iOS was introduced in 2010 with the release of iOS version 4.0.

Fig. 10.4 Android 6.0

10.10   Review Questions

1. What is an operating system?
2. What are the main functions of an operating system?
3. Explain the following operating system concepts: Processor scheduling, multiprogramming, paging/segmentation and multitasking.
4. Explain what is meant by the operating system terms: process, interrupt, concurrency, interleaving, deadlock and livelock.
5. Describe IBM's contributions to operating system development.
6. Describe the similarities and differences between VM and MVS.
7. Describe the influence of the UNIX operating system.
8. Describe the features of DEC's VMS operating system.
9. Describe the operating systems available for touchscreen smartphones.

10.11   Summary

An operating system is a collection of software programs that control the hardware of a computer and makes it usable. It makes the computing power of the hardware available to the users of the computer, and it manages the hardware to achieve good system performance.

The earliest computers did not have an operating system, and the user had exclusive control over a large computer for a specified period. The earliest operating systems were designed in the 1950s with the goal to make more efficient use of the computer (as computers were expensive). These batch-processing systems ran one job at a time, and programs and data were submitted in groups (or batches).

These evolved during the early 1960s into multi-batch systems that were designed to get better utilisation of the expensive computer resources. They could handle several diverse jobs at once. However, software development in this environment was very slow which led operating system designers to develop the concept of multiprogramming in which several jobs are in main memory at once.

IBM announced the System/360 family of computers in 1964, and the computers in the family were designed to use the IBM System/360 operating system (OS/360). UNIX was developed at Bell Labs in the early 1970s, and it is a multitasking and multi-user operating system.

IBM outsourced the development of the operating system for the IBM PC to a small company called Microsoft, and Microsoft had the right to license its operating system, MS/DOS, on IBM compatibles, with PC/DOS (or simply DOS) reserved for IBM personal computers only.

The Macintosh was a paradigm shift in the computer industry, when it was introduced in 1984. Its MAC operating system was GUI based, friendly, intuitive and easy to use. Microsoft Windows is a family of graphical operating systems developed by Microsoft, and it is the dominant operating system on laptops and personal computers. It has failed to make an impact on the smartphone operating system market, which is dominated by Apple's iOS and Google's Android operating systems.

The Android operating system was designed mainly for touchscreen smartphones and tablets, and the iOS operating system is a mobile operating system employed on Apple's mobile devices.

Chapter 11
Overview of Databases

Key Topics

Hierarchical Model
Network Model
Relational Model
Table
Key
Index
SQL
Oracle Database

11.1 Introduction

A database (DB) is essentially an organised collection of data, and it consists of schemas, tables, queries, reports and views. It is organised in such a way that a computer program (termed the database management system) may easily select and analyse the desired pieces of data. A database holds information about many different types of entities, as well as information about the relationships between them.

A database management system (DBMS) is a collection of software programs that allows a user to store, modify and extract data from a database. The interaction between the users and the database is through the DBMS, and it enables the definition, creation, query, update and administration of databases. Historically, there are three main categories of database management systems, which are hierarchical, network and relational models. These differ in how the DBMS organises data internally, which determines the speed and efficiently of data retrieval from the database.

The *network model* database is perceived by the user to be a collection of record types, and relationships between them organised as a network. The network model defines the relationships explicitly as part of the structure of the network. The *hierarchical model* is perceived by a user to be a collection of hierarchies or trees, and it is a more restricted structure than the network model, as only one arrow may enter each box on the network. The *relational model* is perceived by the user to be a collection of tables (or relations), and it has been the most popular category of databases since the 1980s.

© Springer International Publishing AG, part of Springer Nature 2018   217
G. O'Regan, *World of Computing*,
https://doi.org/10.1007/978-3-319-75844-2_11

Early work on database management systems began in the 1960s as part of the Apollo mission to land man on the moon. It was clear that the existing systems were not capable of handling the coordination of the vast amounts of data required for the project. IBM developed the Generalised Update Access Method (GUAM) product in 1964, and this product evolved into Data Language/1 (DL/1). DL/1 is the data management component of the Information Management System (IMS) database, which was one of the earliest database management systems when it was introduced in 1968. IMS used the hierarchical model.

The CODASYL committee[1] set up a database task group and devised a standard, which became known as the 'CODASYL approach'. This became the network standard, and it was defined in the late 1960s and introduced in 1971.

Codd proposed the relational model, a radically new approach to the management of data in 1970, and IBM developed the prototype system called System R in the mid-1970s. Commercial relational database systems were introduced from the early 1980s, and today relational databases are dominant with the network or hierarchical databases mainly of historical interest. Among the popular relational databases used today are Oracle, Microsoft SQL Server and Informix.

11.2 Hierarchical and Network Models

A database management system uses the network model if the data relationships are defined in terms of a graph. The relationships are defined in terms of records (a record is a collection of fields, with each field containing one value), which are connected via links. Any given record may have several parent records and several dependent records, and cycles are permitted in the model. Charles Bachman and others on the CODASYL Committee defined the network model in the late 1960s.

General Electric's integrated data store (IDS), and the integrated database management system (IDMS) are well-known databases that were based on the network model. These mainframe databases were introduced in the early 1970s.

The network model of suppliers and parts allows many-to-many relationships to be expressed, and is presented in a simple graph like structure (Fig. 11.1). There is more detailed information in (Date 1981).

A database management system uses the hierarchical model if the data relationships are defined in terms of hierarchies (i.e. in a tree-like structure). The relationships are simple but inflexible (as they are one to many). The data are defined as records, which are connected to each other through links. Each child record may have only one parent, whereas each parent record may have several

---

[1]The CODASYL committee is the group that defined and standardised the COBOL programming language. It was also involved in work in standardising database interfaces.

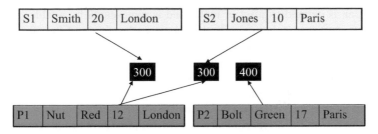

Fig. 11.1 Simple part/supplier—network model

Fig. 11.2 Simple part/
supplier—hierarchical model

children records. The whole tree (starting from the root) needs to be traversed to retrieve data from a hierarchical database. That is, the hierarchical model is a more restricted version of the network model, where no box can have more than one arrow entering the box although several arrows can leave a box.

The hierarchical model of suppliers and parts presents the data in a simple tree-like structure (Fig. 11.2). Each tree consists of one part record together with a set of supplier record occurrences, one for each supplier of the part. There is more detailed information in (Date 1981).

The database access and manipulation component of the hierarchical model is termed Data Language/1, and it includes a data definition language and a data manipulation language. The IBM Information Management System (IMS) is one of the most widely used hierarchical databases, and it was created in the late 1960s.

11.3   The Relational Model

A relational database management system (RDBMS) is a system that manages data using the relational model, and examples include RDMS developed at MIT in the 1970s; Ingres developed at the University of California, Berkeley in the mid-1970s; Oracle developed in the late 1970s; DB2 developed by IBM in the early 1980s; Informix which was originally developed by Informix Corp. in the early 1980s, and acquired by IBM in 2001; and Microsoft SQL Server.

A relation is defined as a set of tuples and is represented by a table. A table is organised in rows and columns, with the data in each column of the table is of the same data type. Constraints may be employed to provide restrictions on the kinds of data that may be stored in the relations. These are Boolean expressions which indicate whether the constraint holds or not, and are a way of implementing business rules in the database.

Relations have one or more keys associated with them, and the *key uniquely identifies the row of the table*. An *index* is a way of providing fast access to the data in a relational database, as it allows the tuple in a relation to be looked up directly (using the index) rather than checking all tuples in the relation.

The structured query language (SQL) is a computer language that tells the relational database what to retrieve and how to display it. A stored procedure is executable code that is associated with the database, and it is used to perform common operations on the database.

The concept of a relational database was first described in a paper '*A Relational Model of Data for Large Shared Data Banks*' by Codd (Codd 1970). A relational database is a database that conforms to the relational model, and it may be defined as a set of relations (or tables).

Codd (Fig. 4.7) was a British mathematician, computer scientist and IBM researcher, who initially worked on the SSEC (selective sequence electronic computer) project in New York, and then on the IBM 701 and 702 computers. He later worked on the IBM 7030 STRETCH computer (IBM's first transistorised computer). He was the creator of STEM (statistical database expert manager).

He developed the *relational database model* in the late 1960s, and he published an internal IBM paper on the relational model in 1969. Today, this is the standard way that information is organised and retrieved from computers, and relational databases are at the heart of systems from hospitals' patient records to airline flight and schedule information.

IBM was promoting its IMS hierarchical database in the 1970s, and it showed little interest or enthusiasm for Codd's new relational database model. It made business sense for IBM to preserve revenue for the IMS/DB model, rather than embarking on a new technology. However, IBM agreed to implement Codd's ideas on the relational model for the *System R research project* in the 1970s, and this project demonstrated the power of the model, as well as demonstrating good transaction processing performance. The project introduced a data query language that was initially called SEQUEL (later renamed to SQL), and this language was designed to retrieve and manipulate data in the IBM database.

Codd continued to develop and extend his relational model, and several theorems are named after him. In later years, he proposed a three-valued logic to deal with missing or undefined information, and he even proposed a four-valued logic in the 1990s. These proposals were never implemented and were controversial at the time. The relational model became popular from the early 1980s, and Codd received the ACM Turing Award in 1981 in recognition of its development.

A binary relation R(A, B) where A and B are sets is a subset of the Cartesian product $(A \times B)$ of A and B. The domain of the relation is A, and the co-domain of the relation is B. The notation aRb signifies that there is a relation between a and b and that $(a,\ b)\in$R. An n-ary relation R $(A_1,\ A_2,\ \ldots,\ A_n)$ is a subset of the Cartesian product of the n sets: i.e. a subset of $(A_1 \times A_2 \times \cdots \times A_n)$. However, an n-ary relation may also be regarded as a binary relation R $(A,\ B)$ with $A = A_1 \times A_2 \times \cdots \times A_{n-1}$ and $B = A_n$.

The data in the relational model are represented as a mathematical n-ary relation. That is, a relation is defined as a set of n-tuples, and a table provides a visual representation of the relation, with the data organised in rows and columns. The data stored in each column of the table is of the same data type.

The basic relational building block is the domain or data type (often called just type). Each row of the table represents one n-tuple (one tuple) of the relation, and the number of tuples in the relation is the cardinality of the relation. Consider the PART relation taken from (Date 1981), where this relation consists of a heading and the body. There are five data types representing part numbers, part names, part colours, part weights and locations in which the parts are stored. The body consists of a set of n-tuples. The PART relation is of cardinality six (Fig. 11.3).

Strictly speaking, there is no ordering defined among the tuples of a relation, since a relation is a set, and a set may be unordered. However, in practice, relations are often considered to have an ordering.

There is a distinction between a domain and the columns (or attributes) that are drawn from that domain. An *attribute* represents the *use* of a domain within a relation, and the distinction is often emphasised by giving attributes names that are distinct from the underlying domain. The difference between domains and attributes can be seen in the PART relation (Fig. 11.4) from Date (1981).

A *normalised relation* satisfies the property that at every row and column position in the table there is exactly one value (i.e. never a set of values). All relations in a relational database are required to satisfy this condition, and an un-normalised relation may be converted into an equivalent normalised form.

It is often the case that within a given relation that there is one attribute with values that is unique within the relation, and can thus be used to identify the tuples of the relation. For example, the attribute P# of the PART relation has this property since each PART tuple contains a distinct P# value, which may be used to distinguish that tuple from all other tuples in the relation. P# is termed the *primary key*

P#	PName	Colour	Weight	City
P1	Nut	Red	12	London
P2	Bolt	Green	17	Paris
P3	Screw	Blue	17	Rome
P4	Screw	Red	14	London
P5	Cam	Blue	12	Paris
P6	Cog	Red	19	London

Fig. 11.3 PART relation

Fig. 11.4 Domains versus attributes	DOMAIN	PART_NUMBER	CHARACTER(6)
	DOMAIN	PART_NAME	CHARACTER(20)
	DOMAIN	COLOUR	CHARACTER(6)
	DOMAIN	WEIGHT	NUMERIC(4)
	DOMAIN	LOCATION	CHARACTER(15)

RELATION PART

	(P#	: DOMAIN PART_NUMBER
	PNAME	: DOMAIN PART_NAME
	COLOUR	: DOMAIN COLOUR
	WEIGHT	: DOMAIN WEIGHT
	CITY	: DOMAIN LOCATION)

for the PART relation, and a candidate key that is not the primary key is termed the *alternate key*.

An index is a way of providing quicker access to the data in a relational database, as it allows the tuple in a relation to be looked up directly (using the index) rather than checking all tuples in the relation.

The consistency of a relational database is enforced by a set of constraints that provide restrictions on the kinds of data that may be stored in the relations. The constraints are declared as part of the logical schema and are enforced by the database management system. They are used to implement the business rules for the database.

11.4   Structured Query Language (SQL)

Codd proposed the Alpha language as the database language for his relational model. However, IBM's implementation of his relational model in the System R project introduced a data query language that was initially called SEQUEL (later renamed to SQL). This language did not adhere to Codd's relational model but it became the most popular and widely used database language. It was designed to retrieve and manipulate data in the IBM database, and its operations include *insert, delete, update, query*, schema creation and modification and data access control.

Structured query language (SQL) is a computer language that tells the relational database what to retrieve and how to display it. It was designed and developed at IBM by Donald Chamberlin and Raymond Boyce, and it became an ISO standard in 1987.

The most common operation in SQL is the query command, which is performed with the SELECT statement. The SELECT statement retrieves data from one or more tables, and the query specifies one or more columns to be included in the result. Consider the example of a query that returns a list of expensive books (defined as books that cost more than $100.00).

SELECT *[2]

> FROM Book
> WHERE Price > 100.00
> ORDER by title;

The *data manipulation language* (DML) is the subset of SQL used to add, update and delete data. It includes the INSERT, UPDATE and DELETE commands. The *data definition language* (DDL) manages table and index structure, and includes the CREATE, ALTER, RENAME and DROP statements.

There are extensions to standard SQL that add programming language functionality. A stored procedure is executable code that is associated with the database. It is usually written in an imperative programming language, and it is used to perform common operations on the database.

Oracle is recognised as a world leader in relational database technology and its products play a key role in business computing. An Oracle database consists of a collection of data managed by an Oracle Database Management System, and Oracle is the main standard for database technology.

11.5 Oracle Database

An Oracle database is a collection of data treated as a unit, and the database is used to store and retrieve related information. The database server manages a large amount of data in a multi-user environment. It allows concurrent access to the data, and the database management system prevents unauthorised access to the database. It also provides a smooth recovery of database information in the case of an outage or any other disruptive event.

Every Oracle database consists of one or more physical data files, which contain the database data, and a control file that contains entries that specify the physical structure of the database.

An Oracle database includes logical storage structures that directly refer to the database's data. A schema is a collection of database objects, which include structures such as tables, views and indexes.

Tables are the basic unit of data storage in an Oracle database, and each table has several rows and columns. An index is an optional structure associated with a table, and it is used to enhance the performance of data retrieval. The index provides an access path to the table data. A view is the customised presentation of data from one or more tables. It does not contain actual data and derives the data from the actual tables on which it is based.

---

[2]The asterisk (*) indicates that all columns of the Book table should be included in the result.

Each Oracle database has a data dictionary, which stores information about the logical and physical structure of the database. The data dictionary is created when the database is created, and is updated automatically by the Oracle database to ensure that it accurately reflects the status of the database.

An Oracle database uses various processes to manage and access the database, including server processes, background processes and user processes. A database administrator (DBA) is responsible for setting up the Oracle database server and application tools, and managing database performance. The role involves allocating system storage and planning future storage requirements for the database management system.

The DBA will create appropriate storage structures to meet the needs of application developers who are designing a new application. The access to the database will be monitored and controlled, and the performance of the database monitored and optimised. The DBA will plan backups and recovery of database information.

11.6   Review Questions

1. What is a database?
2. What is a database management system?
3. Explain relational, hierarchical and network databases.
4. Explain the difference between a key and an index.
5. What is a stored procedure?
6. What is the role of the Oracle DBA?
7. Explain the differences between tables, views and schemas.
8. What is SQL?
9. What is an Oracle database?

11.7   Summary

A database management system is a collection of software programs that allows a user to store, modify and extract data from a database. There are three main categories of database management systems, and these are hierarchical, network and relational models.

A network model database is perceived by the user to be a collection of record types and relationships between them organised as a network. A hierarchical model is perceived by a user to be a collection of hierarchies or trees, and it is a more restricted structure than the network model. A relational model is perceived by the user to be a collection of tables (or relations).

Early work on database management systems began in the 1960s, and IBM developed the information management system (IMS) database in the late 1960s. This hierarchical database was one of the earliest database management systems.

Codd proposed the relational model as a new approach to the management of data in 1970, and IBM developed the prototype System R relational database in the 1970s. Relational databases are now dominant with the hierarchical and network model mainly of historical interest.

Chapter 12
Overview of Telecommunications

Key Topics

Telegraph
Telephone
AMPS
AXE
Telephone
Telegraph
Mobile Phone System
Iridium

12.1 Introduction

Telecommunications is a branch of technology concerned with the transmission of information over a distance, where the transmitter sends the information to a receiver. Early societies used fire and smoke signals for visual communication, with drums used for auditory communication. This allowed simple messages (e.g. 'danger') to be communicated to other groups.

The Persian Empire established an early postal system in the sixth century B.C., and the Egyptians and Romans later established their own postal systems. A pigeon messaging system, where the homing characteristics of pigeons were employed to send messages, was later introduced.

The Greeks introduced an early semaphore system in the fourth century B.C., which allowed very simple messages to be exchanged between groups on two different hills (similar in a sense to smoke signals). A ship semaphore system was introduced in the fifteenth century, which allowed two ships to communicate with each other. This system used flags where the position and motion of a flag represented a letter.

The Chappe brothers in France introduced an early optical telegraph system in Europe in the late eighteenth century. It used similar principles as the ship-based semaphore system, and it allowed messages to be sent from one high tower to another. It was used by the French military.

© Springer International Publishing AG, part of Springer Nature 2018
G. O'Regan, *World of Computing*,
https://doi.org/10.1007/978-3-319-75844-2_12

Early electrical telegraph systems were introduced in the early nineteenth century, and Samuel Morse devised a system (the Morse code) that allowed letters to be represented by a series of on–off tones in the late 1830s. This was the foundation for electrical telegraphs and later telephone systems. The first Atlantic telegraph cable was laid between Britain and America (via Valencia Island in Ireland) in 1858, and this allowed messages to be sent and responded to the same day rather than the usual delivery time of 10 days for letters sent by ships.

The telephone was invented by Alexander Graham Bell in 1876[1], and early telephones were hardwired to and communicated with a single other telephone (e.g. from a person's business to his home), as initially there were no telephone exchanges. A telephone exchange provides switching or interconnection between two subscriber lines, and the earliest manual commercial telephone exchanges were introduced in the late 1870s. The first mechanical automated exchanges were introduced in the early 1890s. The first North American transcontinental phone call from the east coast to the west coast was made by Bell in 1915, and it made long-distance communication a reality.

The invention of the telephone was a paradigm shift from *face-to-face* communication, where people met to exchange ideas and share information, or where individuals wrote letters to each other to exchange information. The telephone was a new medium that provided direct and instantaneous communication between two people. It allowed two individuals to establish and maintain two-way communication irrespective of being at two different physical locations. Initially, the business community and the affluent members of society used the telephone, but this changed rapidly in the years that followed.

Marconi, an Italian engineer, introduced a system for the wireless transmission of sounds in 1896, and the British Marconi Company was established in 1897. It began communication between ships at sea and coastal radio stations, and the first radio messages were sent across the Atlantic in 1902. The awareness of the value of radio communication was highlighted in the sinking of the Titanic in 1912. Marconi established an early radio factory in England in 1912.

The first prototype electronic television was developed and demonstrated by Philip Farnsworth in the late 1920s. It was the result of research on ways to transmit images, and it had been determined that radio waves could be encoded with an image, and then transmitted back to the screen. Farnsworth's prototype is considered the first electronic television.

The foundations of the mobile cellular industry go back to the introduction of a limited-capacity mobile phone system that was introduced for automobiles in 1946. Martin Cooper of Motorola made the first mobile phone call to Joe Engels at Bell Labs in 1973, and a prototype mobile phone network was operational in the late 1970s with commercial mobile phone networks introduced in the early 1980s. The

---

[1]He was the first person to patent the telephone as an 'apparatus for transmitting vocal or other sounds telegraphically'. There are several other claimants for inventing the telephone.

first global mobile phone system (Iridium) was operational in 1998, and the Iridium system consisted of 66 satellites, with the customers using handheld satellite phones.

The ARPANET packet switching network was introduced in the late 1960s, and it remained operational until 1990, when the Internet became operational. The Internet has led to almost instantaneous communication, and it has led to electronic mail, the World Wide Web, which was developed by Tim Berners Lee at CERN, social networking, electronic commerce and telephone calls over the Internet with the VOIP protocol.

This chapter considers a small number of events in the history of telecommunications including the development of the AXE system, which was the first fully automated digital switching system, the development of mobile phone technology and the development of the Iridium satellite mobile phone system.

12.2 AXE System

Ericsson introduced the AXE (Automatic Exchange Electric) switching system in 1977 (Fig. 12.1). This was the first fully automated digital switching system, and it converted speech into digital (i.e. the binary language used by computers). Ericsson's competitors were still using the slower and less reliable analog systems.

The analog system uses an electric current to convey the vibrations of the human voice, whereas a digital system uses a stream of binary digits to represent sound. The AXE system was an immediate success with telecom companies, and it has

Fig. 12.1 AXE system.
Courtesy of Ericsson

been sold in many countries around the world. AXE was originally a digital exchange for landline telephony, but it was later extended for use with mobile telephony systems.

Ellemtel was established in 1970 as a pure research and development company, and it was a joint venture between Televerket (Sweden's state-owned PTT) and Ericsson. Its primary task was to develop an electronic and automated switching system for telephone stations that would become the AXE system.

Ericsson had been working to develop a commercial electronic switching system called AKE, while Televerket was working on its own electronic switch. Ericsson realised that its AKE system was not suitable for large switching stations, and that it needed to develop a new generation of switching systems. It decided to combine its resources with Televerket and to jointly develop an electronic telephone switching system.

Bengt-Gunnar Magnusson was the project manager for the AXE project, and AXE had a modular system design which made the system flexible. New functionality could be added, and existing modules updated or replaced. The modular design allowed the system to be easily adapted to different markets.

The development of AXE also involved the development of hardware and software such as programs and processors to control the AXE stations. The first prototype AXE system was installed at a Televerket station in 1976, and Ellemtel's work in developing the AXE system was complete in 1978.

The AXE system was then commercialised and many of Ellemtel's employees moved to Ericsson. AXE was an immediate success and Ericsson soon had customers in Sweden, Finland, France, Australia and Saudi Arabia. The Saudi order was the largest that Ericsson had ever received, and it involved increasing the capacity of the Saudi network by 200% and installing the AXE system.

The introduction of AXE meant that by the early 1980s that Ericsson had the market's most advanced and flexible switching system, and this made it ideally placed for the transition from fixed line to mobile telephony. It meant that Ericsson had moved from being a minor player in the telecoms business to a major-league player. It was now the leader in fixed line phone technology, and it had the right foundations in place for success in mobile telephony. It became the leader in mobile technology in the late 1980s, and today the AXE system has been installed in over 130 countries.

12.3 Development of Mobile Phone Standards

Bell Labs played an important role (with Motorola) in the development of the analog mobile phone system in the United States. It developed a system in the mid-1940s that allowed mobile users to place and receive calls from automobiles, and Motorola developed mobile phones for automobiles. However, these phones were large and bulky and they consumed a lot of power. A user needed to keep the automobile's engine running to make or receive a call.

Fig. 12.2 Frequency reuse in cellular networks

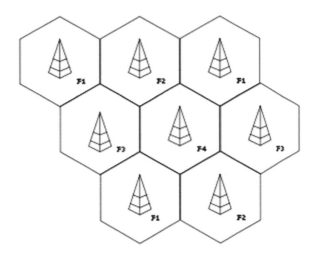

Bell Labs first proposed the idea of a cellular system back in the late 1940s, when they proposed hexagonal rings for mobile communication. Large geographical areas were divided into cells, where each cell had its own base station and channels. The available frequencies could be used in parallel in different cells without disturbing each other (Fig. 12.2). Mobile telephone could now, in theory, handle many subscribers. However, it was not until the late 1960s that Bell Labs prepared a detailed plan for implementing the cellular system.

Bell Labs developed the Advanced Mobile Phone System (AMPS) standard from 1968 to 1983. Motorola and other telecommunication companies designed and built phones for this cellular system. The AMPS system uses separate frequencies (or channels) for each conversation and requires considerable bandwidth for many users.

The signals from a transmitter cover an area called a cell. As a user moves from one cell into a new cell, a handover to the new cell takes place without any noticeable difference to the user. The signals in the adjacent cell are sent and received on different channels to the existing cell's signals, and so there is no interference.

The Total Access Communication (TACS) and extended TACS (ETACS) system were variants of AMPS that were employed in the United Kingdom and Europe. These analog standards employed separate frequencies (or channels) for each conversation using frequency division multiple access (FDMA). However, the analog system suffered from static and noise, and there was no protection from eavesdropping using a scanner.

Ericsson became the leader in the first generation of mobile with Motorola, and the extent of its leadership was clear when its proposed design for digital mobile radio transmission was selected as the US Standard for Cellular Communications over entries from Motorola and AT&T in 1989.

The AMPS system represents the first generation of cellular technology, and it has several weaknesses when compared to today's cellular systems. Mobile technology evolved to the second-generation digital global system for mobile communication (GSM) and code division multiple access (CDMA) technologies; to general packet radio service (GPRS); to third-generation mobile, including 3G and WCDMA; and to fourth- and fifth-generation mobile (4G and 5G).

12.4 Development of Mobile Phone Technology

The invention of the telephone by Graham Bell in the late nineteenth century was a revolution in human communication, as it allowed people in different geographic locations to communicate instantaneously rather than meeting face-to-face. However, the key restriction of the telephone was that the actual physical location of the person to be contacted was required prior to communication, as otherwise communication could not take place, i.e. *communication was between places rather than people*.

The origins of the mobile phone revolution date to work done on radio technology in the 1940s. Bell Labs had proposed the idea of a cellular communication system back in 1947, and it was eventually brought to fruition by researchers at Bell Labs and Motorola. Bell Labs constructed and operated a prototype cellular system in Chicago in the late 1970s, and performed public trials in 1979. Motorola commenced the second U.S. cellular system test in the Washington/Baltimore area. The first commercial systems commenced operation in the United States in 1983.

The DynaTAC (Dynamic adoptive Total Area Coverage) used cellular radio technology to link people and not places. Motorola was the first company to incorporate the technology into a portable device designed for use outside of an automobile, and it spent $100 million on the development of cellular technology. Martin Cooper (Fig. 12.3) led the team at Motorola that developed the DynaTAC8000X, and he made the first mobile phone call on a prototype DynaTAC phone to Joel Engels, the head of research at Bell Labs, in April 1973.

Commercial cellular services commenced in North America in 1983, and the world's first commercial mobile phone went on sale the same year. This was the Motorola DynaTAC 8000X, and it was popularly known as the '*brick*' due to its size and shape. It weighed 28 oz (almost 2 lbs); it was 13.5" (over a foot) in length and 3.5" in width. It had a LED display and could store 30 numbers. It had a talk time of 30 min, 8 h of standby, and it took over 10 h to recharge.

The cost of the Motorola DynacTAC 8000X was $3995, and it was too expensive for most people apart from wealthy consumers. Today, mobile phones are ubiquitous, and there are more mobile phone users than fixed line users. The cost of a mobile phone today is typically less than $100, and a mobile phone typically weighs as little as 3 oz.

The first-generation mobile phone system introduced into North America in the early 1980s used the 800 MHz cellular band. It had a frequency range between 800

Fig. 12.3 Martin Cooper
re-enacts DynaTAC call

and 900 MHz. Each service provider could use half of the 824–849 MHz range for receiving signals from cellular phones and half the 869–894 MHz range for transmitting to cellular phones. The bands were divided into 30 kHz sub-bands called channels and a separate frequency (or channel) was used for each conversation. The division of the spectrum into sub-band channels is achieved by using frequency division multiple access (FDMA).

This first-generation system allowed voice communication only, and it was susceptible to static and noise. Further, it had no protection from eavesdropping using a scanner.

The AXE system provided the foundation for Ericsson's growth in mobile telephony, as its flexible modular design allowed new functionality to be added, and by changing a module AXE could be reconfigured to handle mobile telephone calls. This allowed Ericsson to design the first mobile telephone exchange (MTX) by replacing the sub-system for fixed subscribers with a new sub-system for mobile subscribers. The MTX switch was developed in the late 1970s/early 1980s and was a key part of the Nordic mobile telephone system (NMT) which would be used in all Nordic countries.

Ericsson was awarded a large Saudi Arabian contract to deliver a fixed line and mobile system, and it was agreed that the NMT standard would be used and that

Ericsson would supply the entire system. The Saudi mobile phone network became operational from 1981, and Ericsson provided base stations, radio towers and switches. Ericsson had now acquired cell-planning experience, and it was awarded the contract to develop the entire mobile telephone network in the Netherlands. Ericsson was now a total systems supplier in mobile telephony, and it provided the entire infrastructure such as switches and base stations. Today, its base stations range from small picocells to large macrocells.

The second generation (2G) of mobile technology was a significant improvement on the existing analog technology. This digital, cellular technology encrypted telephone conversations and provided data services such as text and picture messages. The second-generation technologies included the GSM standard developed by the European Telecommunications Standards Institute (ETSI), and CDMA developed in the United States. The first GSM call was made by the Finnish prime minister in Finland in 1991, and the first short message service (SMS) or text message was sent in 1992.

The subscriber identity module (SIM) card was a new feature in GSM, and a SIM card is a detachable smart card that contains the user's subscription information and phone book. The SIM card may be used in other GSM phones, and this is useful when the user purchases a replacement phone. GSM provides an increased level of security, with communication between the subscriber and base station encrypted.

GSM networks evolved into GPRS (2.5 G), which became available in 2000. Third- and fourth-generation mobile (3G and 4G) provide mobile broadband multimedia communication. Mobile phone technology has transformed the earlier paradigm of *communication between places* to that of *communication between people*.

Motorola dominated the analog mobile phone market. However, it was slow to adapt to the GSM standard, and it paid a heavy price with a loss of market share to Nokia and Ericsson. It was very slow to see the potential of a mobile phone as a fashion device,[2] and it was too slow in adapting to smartphones.

12.5   The Iridium Satellite System

Iridium was a global satellite phone company that was backed by Motorola. In many ways, it was an engineering triumph over common sense, and over \$5 billion was spent in building an infrastructure of low earth orbit (LEO) satellites to provide global coverage. It was launched in late 1998 to provide worldwide wireless coverage to its customers, including the oceans, airways and polar regions. The existing telecom systems had limited coverage in remote areas, and so the concept of global coverage as provided by Iridium was potentially very useful.

---

[2]The attitude of Motorola at the time seemed to be like that of Henry Ford, i.e. they can have whatever colour they like as long as it is black.

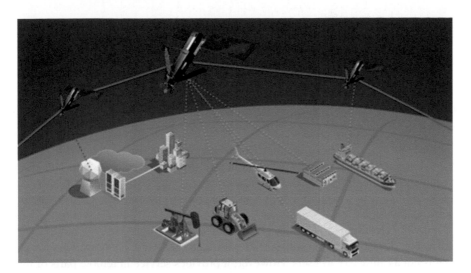

Fig. 12.4 Iridium system. Courtesy of Iridium Satellite LLC

Iridium was implemented by a constellation of 66 satellites (Fig. 12.4). The original design required 77 satellites, and so the name '*Iridium*' was chosen (since its atomic number in the periodic table is 77). However, the later design required just 66 satellites, and so '*Dysprosium*' may have been a more appropriate choice. The satellites are in low earth orbit at a height of approximately 485 miles, and communication between the satellites is via inter-satellite links. Each satellite contains seven Motorola Power PC 603E processors running at 200 MHz, which are used for satellite communication and control.

Iridium routes phone calls through space and there are several earth stations. As satellites leave the area of an earth base station the routing tables change, and frames are forwarded to the next satellite just coming into view of the earth base station.

The Iridium constellation is a large commercial satellite constellation, and it is especially suited for industries such as maritime, aviation, government and the military. Motorola was the prime contractor for Iridium, and it played a key role in its design and development. The satellites were produced at a cost of $5 million each ($40 million each including launch costs), and Motorola engineers could make a satellite in the phenomenal time of 2–3 weeks.

The first Iridium call was made by Al Gore in late 1998. However, despite being an engineering triumph, Iridium was a commercial failure, and it went bankrupt in late 1999 due to insufficient demand for its services. It had needed a million subscribers to break even, and as the cost of an Iridium call was very expensive compared to the existing cellular providers, and as the cost of its handsets was much higher and more cumbersome to use than existing mobile phones, there was very little demand for its services.

Specifically, the reasons for failure included:

- Insufficient demand for its services (10,000 subscribers),
- High cost of its service ($5 per minute for a call),
- Cost of its mobile handsets ($3000 per handset),
- Bulky mobile handsets,
- Competition from existing mobile phone networks and
- Management failures.

However, the Iridium satellites remained in orbit, and the service was re-established in 2001 by the newly founded Iridium satellite LLC. The new business model required just 60,000 subscribers to break even. Today, it has over half a million customers, and it is used extensively by the US Department of Defense.

Iridium was designed in the late 1980s and so it is designed primarily for voice rather than data. It lacks the sophistication of modern mobile phone networks, and so it is not as attractive to users. However, it provides service in remote parts of the world, which is very useful. Iridium is developing and launching a second generation of satellites (Iridium Next), which will include new features such as data transmission.

12.6 Review Questions

1. Describe the contributions of Bell Labs to mobile technology.
2. What are the advantages of mobile technology over fixed line technology?
3. Describe the various generations of mobile technology.
4. Describe Motorola's contributions to mobile technology.
5. What factors led to Ericsson's success and leadership in mobile technology?
6. What factors led to the (initial) commercial failure of the Iridium System?

12.7 Summary

The invention of the telephone by Graham Bell in the late nineteenth century was a revolution in human communication, as it allowed people in different geographic locations to communicate instantaneously rather than meeting face-to-face. The early phones had major limitations, but the development of automated telephone exchanges helped to deal with these.

However, the key limitation of the telephone was that the actual physical location of the person to be contacted was needed prior to communication, i.e. communication was between places rather than people.

This led to research by Bell Labs and others into ways in which communication could take place between people (and not places). Bell Labs developed a system in the mid-1940s that allowed mobile users to place and receive calls from automobiles, with Motorola developing the phones for automobiles. However, these phones were large and bulky, and the automobile's engine needed to be running to make or receive a call.

Bell Labs proposed the idea of a cellular system back in the late 1940s, and it prepared a detailed plan for its implementation in the late 1960s. A cellular system is divided into cells, where each cell has its own base station and channels. The available frequencies may be used in parallel in different cells without interference with each other.

Motorola developed the first mobile phone, the DynaTAC, and it made the first mobile phone call in 1973. The first mobile phone systems were analog and based on the AMPS standard. The later generations of mobile technology are digital and are a significant advance on the older cellular technology.

Iridium provides global wireless coverage to its customers including coverage in the oceans, airways and polar regions. It was implemented by a constellation of 66 satellites. For a more detailed account of the contributions of Bell Labs, Ericsson and Motorola, see (Gertner 2013; Meurling and Jeans 2001; Motorola 1999; O'Regan 2015).

Chapter 13
The Internet and World Wide Web

Key Topics

ARPANET
TCP/IP
The Internet
Internet of Things
Internet of Money
The world Wide Web
Dot-Com Bubble
Facebook
The Twitter Revolution

13.1 Introduction

The vision of the Internet and World Wide Web goes back to an article by Vannevar Bush in the 1940s. Bush was an American scientist who had done work on submarine detection for the U.S. Navy. He designed and developed the Differential Analyser (Fig. 1.1), which was a mechanical computer whose function was to evaluate and solve first-order differential equations. Bush supervised Claude Shannon at MIT (see Chap. 2), and Shannon's initial work was to improve the Differential Analyser.

Bush (Fig. 13.1) became director of the office of Scientific Research and Development, and he developed a win-win relationship between the U.S. military and universities. He arranged generous research funding for the universities to carry

© Springer International Publishing AG, part of Springer Nature 2018
G. O'Regan, *World of Computing*,
https://doi.org/10.1007/978-3-319-75844-2_13

Fig. 13.1 Vannevar Bush

out applied research to assist the military. This allowed the military to benefit from the early exploitation of research results, and it also led to better facilities and laboratories at the universities. It led to close links and cooperation between universities such as Harvard and Berkeley, and this would eventually lead to the development of ARPANET by DARPA.

Bush outlined his vision of an information management system called the '*memex*' (memory extender) in a famous essay '*As We May Think*' (Bush 1945). He envisaged the memex as a device electronically linked to a library that would be able to display books and films. It describes a proto-hypertext computer system and influences the later development of hypertext systems.

> *A memex is a device in which an individual stores all his books, records, and communications, and which is mechanized so that it may be consulted with exceeding speed and flexibility. It is an enlarged intimate supplement to his memory.*
>
> *It consists of a desk, and while it can presumably be operated from a distance, it is primarily the piece of furniture at which he works. On the top are slanting translucent screens, on which material can be projected for convenient reading. There is a keyboard, and sets of buttons and levers. Otherwise it looks like an ordinary desk.*

Bush predicted that:

> *Wholly new forms of encyclopedias will appear, ready made with a mesh of associative trails running through them, ready to be dropped into the memex and there amplified.*

This description motivated Ted Nelson and Douglas Engelbart to independently formulate ideas that would become hypertext. Tim Berners-Lee would later use hypertext as part of the development of the World Wide Web.

13.2 The ARPANET

There were approximately 10,000 computers in the world in the 1960s. These were expensive machines (often over $1 million) with limited processing power. They contained only a few thousand words of magnetic memory, and programming and debugging was difficult. Further, communication between computers was virtually non-existent.

However, several computer scientists had dreams of worldwide networks of computers, where every computer around the globe is interconnected to all other computers in the world. Licklider[1] wrote memos in the early 1960s on his concept of an intergalactic network, which envisaged that everyone around the globe would be interconnected and able to access programs and data at any site from anywhere.

The U.S. Department of Defense founded the Advance Research Projects Agency (ARPA) in the late 1950s. ARPA embraced high-risk, high-return research, and Licklider became the head of its computer research program. He developed close links with MIT, UCLA and BBN Technologies.[2] The concept of packet switching[3] was invented in the 1960s, and several organisations including the National Physical Laboratory (NPL), the RAND Corporation and MIT commenced work on its implementation.

The early computers had different standards for data representation, and so it was essential to know the standard employed by each computer prior to communication. This led to recognition of the need for common standards in data representation, and a U.S. government committee developed the American Standard Code for Information Interchange (ASCII) in 1963. This was the first universal standard for data, and it allowed machines from different manufacturers to exchange data. The standard allowed a 7-bit binary number to stand for a letter in the English alphabet, an Arabic numeral or a punctuation symbol. The use of 7 bits allowed 128 distinct characters to be represented. The development of the IBM System/360 mainframe (discussed in Chap. 3) standardised the use of 8-bits for a word, and 12-bit or 36-bit words became obsolete.

We discussed the SAGE system and early work done on wide-area networks for military use in the late 1950s (see Chap. 3). The first civilian wide-area network connection was created in 1965, and it involved the connection of a computer at MIT to a computer in Santa Monica. This was done via a dedicated telephone line, and it showed that a telephone line could be used for data transfer. ARPA

---

[1]Licklider was an early pioneer of AI and wrote an influential paper 'Man-Computer Symbiosis' in 1960 (Licklider 1960), which outlined the need for simple interaction between users and computers.

[2]BBN Technologies (originally Bolt Beranek and Newman) is a research and development technology company. It played an important role in the development of packet switching and in the implementation and operation of ARPANET. The '@' sign used in an email address was a BBN innovation.

[3]Packet switching is a message communication system between computers. Long messages are split into packets, which are then sent separately to minimise the risk of congestion.

recognised the need to build a network of computers, and this led to the ARPANET project in 1966 which aimed to implement a packet-switched network with a network speed of 56 Kbps. ARPANET was to become the world's first packet-switched network.

BBN Technologies was awarded the contract to implement the network, with plans for a total of 19 nodes. The first two nodes were based at UCLA and Stanford Research Institute (SRI). The network management was performed by interconnected *Interface Message Processors* (IMPs), which were in front of the main computers. The IMPs eventually evolved to become the network routers that are used today.

The team at UCLA called itself the *Network Working Group*, and it saw its role as developing a set of rules that specified how the computers on the network should communicate. These rules were called the *Network Control Protocol* (NCP). The first host-to-host connection was made between a computer in UCLA and a computer at SRI in late 1969. Several other nodes were added to the network until it reached its target of 19 nodes in 1971.

The Network Working Group developed the *telnet protocol* and the *file transfer protocol* (FTP) in 1971. The telnet program allowed the user of one computer to remotely log into the computer of another computer. The file transfer protocol allows the user of one computer to send (or receive) files to (from) another computer. A highly successful public demonstration of ARPANET was made in 1972, and one of the earliest demos was that of Weizenbaum's ELIZA program (see O' Regan 2016). This famous artificial intelligence (AI) program allowed a user to conduct a typed conversation with an artificially intelligent machine (Rogerian psychotherapist) at MIT.

The viability of packet switching as a standard for network communication had been clearly demonstrated. Ray Tomlinson of BBN Technologies developed a program that allowed electronic mail to be sent over the ARPANET. Over 30 institutions were connected to the ARPANET by the early 1970s.

13.3   TCP/IP

ARPA was renamed to the Defense Advanced Research Projects Agency (DARPA) in 1973. It commenced a project to connect seven computers on four islands using a radio-based network, and a project to establish a satellite connection between a site in Norway and in the United Kingdom. This led to a need for the interconnection of the ARPANET with other networks. The key problems were to investigate ways of achieving convergence between ARPANET, radio-based networks and the satellite networks, as these all had different interfaces, packet sizes and transmission rates. Therefore, there was a need for a *network-to-network connection protocol*.

An international network working group (INWG) was formed in 1973. The concept of the transmission control protocol (TCP) was developed at DARPA by Bob Kahn and Vint Cerf, and they presented their ideas at an INWG meeting at the

University of Sussex in England in 1974 (Kahn and Cerf 1974). TCP allowed cross-network connections, and it began to replace the original NCP protocol that was used in ARPANET.

TCP is a set of network standards that specify the details of how computers communicate, as well as the standards for interconnecting networks and computers. It was designed to be flexible and provides a transmission standard that deals with physical differences in host computers, routers and networks. It is designed to transfer data over networks which support different packet sizes, and which may sometimes lose packets. It allows the inter-networking of very different networks, which then act as one network.

The new protocol standards were known as the *transport control protocol* (TCP) and the *Internet protocol* (IP). TCP details how information is broken into packets and re-assembled on delivery, whereas IP is focused on sending the packet across the network. These standards allow users to send electronic mail or to transfer files electronically, without needing to concern themselves with the physical differences in the networks. TCP/IP consists of four layers (Table 13.1).

The Internet protocol (IP) is a connectionless protocol that is responsible for addressing and routing packets. It breaks large packets down into smaller packets when they are travelling through a network that supports smaller packets. A *connectionless protocol means that a session is not established before data is exchanged*, and packet delivery with IP is not guaranteed, as packets may be lost or delivered out of sequence.

An acknowledgement is not sent when data is received, and the sender or receiver is not informed when a packet is lost or delivered out of sequence. The router forwards a packet only if it knows a route to the destination, and otherwise the packet is dropped. Packets are dropped if their checksum is invalid or if their time to live is zero. The acknowledgement of packets is the responsibility of the TCP protocol. The ARPANET employed the TCP/IP protocols as a standard from 1983.

Table 13.1 TCP layers

Layer	Description
Network interface layer	This layer is responsible for formatting packets and placing them on to the underlying network
Internet layer	This layer is responsible for network addressing. It includes the Internet protocol and the address resolution protocol
Transport layer	This layer is concerned with data transport and is implemented by TCP and the user datagram protocol (UDP)
Application layer	This layer is responsible for liaising between user applications and the transport layer. It includes the file transfer protocol (FTP), telnet, domain naming system (DNS) and simple mail transfer program (SMTP)

13.4 Birth of the Internet

The use of ARPANET was initially limited to academia and to the United States military, and in the early years there was little interest from industrial companies. It allowed messages to be sent between the universities that were part of ARPANET. There were over 2000 hosts on the TCP/IP enabled network by the mid-1980s.

It was decided to shut down the network by the late-1980s, and the National Science Foundation (NSF) commenced work on its successor, the NSFNET, in the mid-1980s. This network consisted of multiple regional networks connected to a major backbone. The original links in NSFNET were 56 Kbps but these were updated to 1.544 Mbps T1 links in 1988. The NSFNET T1 backbone initially connected 13 sites, but this increased, as there was growing academic and industrial interest from around the world. The NSF quickly realised that the Internet had commercial potential.

The Internet began to become more international with nodes in Canada and several European countries. DARPA formed the computer emergency response team (CERT) to deal with any emergency incidents arising from the operation of the network.

The independent not-for-profit company, advanced network services (ANS), was founded in 1991. It installed a new network (ANSNET) that replaced the NSFNET T1 network, and it operated over T3 (45 Mbps) links. It was owned and operated by a private company rather than the U.S. government, with the NSF focusing on the research aspects of networks rather than on the operational side.

The ANSNET network was a distributive network architecture operated by commercial providers such as Sprint, MCI and BBN. The various parts of the network were connected by major network exchange points. These were termed network access points (NAPs), and there were over 160,000 hosts connected to the Internet by the late 1980s.

13.5 Birth of the World Wide Web

Tim Berners-Lee invented the World Wide Web at CERN in 1990 (Berners-Lee 2000). CERN is an important European centre for research in the nuclear field, and it is based in Switzerland. It employs several thousand physicists and scientists from around the world, and has many visiting scientists.

One of the problems that scientists at CERN faced in the late 1980s was in keeping track of people, computers, documents and databases. The centre had many visiting scientists who spent several months there, as well as a large pool of permanent staff. There was no efficient way in CERN at that time to share information among scientists.

A visiting scientist might need to obtain information or data from a CERN computer, or to make the results of their research available to researchers at CERN.

Berners-Lee came to CERN in the early 1980s, and he developed a program called 'Enquire' to assist with information sharing and in keeping track of the work of visiting scientists. He returned to CERN in the mid-1980s to work on other projects, and he devoted part of his free time to consider solutions to the information-sharing problem.

He built on several existing inventions such as the Internet, hypertext and the mouse. Ted Nelson invented hypertext in the 1960s, and it allowed links to be present in text. For example, a document such as a book contains a table of contents, an index and a bibliography. These are all links to material that is either within the book itself or external to the book. The reader of a book may follow the link to obtain the internal or external information. Doug Engelbart invented the mouse in the 1960s, and it allowed the cursor to be steered around the screen.

The major leap that Berners-Lee made was essentially a marriage of the Internet, hypertext and the mouse into what has become the World Wide Web. His vision and its subsequent realisation benefited CERN and the wider world.

He created a system that gives every web page a standard address called the universal resource locator (URL). Each page is accessible via the hypertext transfer protocol (HTTP), and the page is formatted with the hypertext mark-up language (HTML). Each page is visible using a web browser. The key features of Berners-Lee invention are listed in Table 13.2.

Berners-Lee invented the well-known terms such as URL, HTML and World Wide Web, and he wrote the first browser program that allowed users to access web pages throughout the world. Browsers are used to connect to remote computers over the Internet, and to request, retrieve and display the web pages on the local machine.

The early browsers included Gopher developed at the University of Minnesota, and Mosaic developed at the University of Illinois. These were replaced in later years by Netscape, which dominated the browser market until Microsoft developed Internet Explorer (IE). The development of the graphical browsers led to the commercialisation of the World Wide Web.

The World Wide Web creates a space in which users can access information easily from any part of the world. This is done using only a web browser and simple web addresses. The user can then click on hyperlinks on web pages to access further

Table 13.2 Features of World Wide Web

Feature	Description
URL	Universal resource identifier [later renamed to Universal Resource Locator (URL)] provides a unique address code for each web page
HTML	Hypertext mark-up language (HTML) is used for designing the layout of web pages
HTTP	The hypertext transport protocol (HTTP) allows a new web page to be accessed from the current page
Browser	A browser is a client program that allows a user to interact with the pages and information on the World Wide Web

relevant information that may be on an entirely different continent. Berners-Lee later became the director of the World Wide Web Consortium, and this MIT-based organisation sets the software standards for the web.

The invention of the World Wide Web was a revolutionary milestone in the history of computing. It transformed the use of the Internet from mainly academic use to where it is now an integral part of peoples' lives. Users may now surf the web, i.e. hyperlink among the millions of computers in the world and obtain information easily. It is revolutionary in that:

- No single organisation is controlling the web.
- No single computer is controlling the web.
- Millions of computers are interconnected.
- It is an enormous market place of billions of users.
- The web is not located in one physical location.
- The web is a space and not a physical thing.

13.6   Applications of the World Wide Web

Berners-Lee realised that the World Wide Web offered the potential to conduct business in cyberspace, rather than the traditional way where buyers and sellers come together to do business in the marketplace.

> *Anyone can trade with anyone else except that they do not have to go to the market square to do so*

The growth of the World Wide Web has been phenomenal, with exponential growth rate curves a feature of newly formed Internet companies and their business plans. It has been applied to many areas including:

- Travel industry (Booking flights, train tickets and hotels),
- E-Marketing,
- Online shopping,
- Portal sites,
- Recruitment services,
- Internet banking,
- Online casinos,
- Online auction sites,
- Newspapers and news channels and
- Social media.

The prediction in the early days was that the new web-based economy would replace traditional bricks and mortar companies. It was expected that most business would be conducted over the web, with traditional enterprises losing market share and going out of business. Exponential growth of e-commerce companies was

Table 13.3 Characteristics of e-commerce

Feature	Description
Catalogue of products	The catalogue of products details the products available for sale and their prices
Well-designed and easy to use	This is essential as otherwise the website will not be used
Shopping carts	This is analogous to shopping carts in a supermarket
Security	Security of credit card information is a key concern for users of the web, as users need to have confidence that their credit card details will not be compromised
Payments	Once the user has completed the selection of purchases, there is a checkout facility to arrange for the purchase of the goods
Order fulfilment/ order enquiry	Once payment has been received, the products must be delivered to the customer

predicted, and the size of the new web economy was estimated to be in trillions of U.S. dollars.

New companies were formed to exploit the opportunities of the web, and existing companies developed e-business and e-commerce strategies to adapt to the brave new world. Companies providing full e-commerce solutions were concerned with the selling of products or services over the web to either businesses or consumers. These business models are referred to as Business-to-Business (B2B) or Business-to-consumer (B2C). E-commerce websites have the following characteristics (Table 13.3).

13.7   Dot-Com Companies

The success of the World Wide Web was phenomenal and it led to a boom in the formation of '*new economy*' businesses. These businesses were conducted over the web and included the Internet portal company, Yahoo; the online bookstore, Amazon; and the online auction site, eBay. Yahoo provides news and a range of services, and most of its revenue comes from advertisements. Amazon initially sold books, but it now sells a collection of consumer and electronic goods. eBay brings buyers and sellers together in an online auction space.

Some of these new technology companies were successful and remain in business. Others were financial disasters due to poor business models, poor management and poor implementation of the new technology. Some of these technology companies offered an Internet version of traditional bricks and mortar company, with others providing a unique business offering. For example, eBay offers an auctioneering Internet site to consumers worldwide which was a totally new service and quite distinct from traditional auctioneering.

David Filo and Jerry Yang founded Yahoo, and they used it to keep track of their personal interests and the corresponding websites on the Internet. Filo and Yang were students at Stanford in California, and their list of interests grew over time and became too long and unwieldy. Therefore, they broke their interests into a set of categories and then subcategories, and this is the core concept of the website.

There was a lot of interest in the site from other students, family and friends and a growing community of users. The founders realised that the site had commercial potential, and they incorporated it as a business in 1995. The company launched its initial public offering (IPO) 1 year later in April 1996, and it was valued at $850 million. Yahoo is a portal site and it offers free email accounts to users, a search engine, news, shopping, entertainment, health and so on. The company earns most of its revenue from advertisement (including the click-through advertisements that appear on a Yahoo web page).

Jeff Bezos founded Amazon in 1995 as an online bookstore, and its product portfolio has expanded to include just about everything. Its initial focus was to build up the 'Amazon' brand throughout the world, and to become the world's largest bookstore. It initially sold books at a loss by giving discounts to buyers to build market share. It was very effective in building its brand through advertisements, marketing and discounts.

It became the largest online bookstore in the world and has a solid business model with a very large product catalogue, a well-designed website with good searching facilities, good check out facilities and good order fulfilment. It also developed an associate model, which allows its associates to receive a commission for purchases of Amazon products made through the associate site.

Pierre Omidyar founded eBay in 1995, and the site brings buyers and sellers together. Millions of items are listed, bought and sold on eBay every day. The sellers are individuals or international companies. Any legal product that does not violate the company's terms of service may be bought or sold on the site. A buyer makes a bid for a product or service, and competes against several other bidders. The highest bid is successful, and payment and delivery are then arranged. The revenue earned by eBay includes fees to list a product and commission fees that are applied whenever a product is sold.

Any product listed that violates eBay's terms of service is removed from the site as soon as the company is aware of them. The company also has a fraud prevention mechanism, which allows buyers and sellers to provide feedback on each other and to rate each other following the transaction. The feedback may be positive, negative or neutral, and relevant comments included. This offers a way to help to reduce fraud as unscrupulous sellers or buyers will receive negative ratings and comments.

13.7.1   Dot-Com Failures

Several of the companies formed during the dot-com era were successful and remain in business today. Others had inappropriate business models or poor management and failed in a spectacular fashion. This section considers some of the dot-com failures and highlights the reasons for failure.

Webvan.com was an online grocery business based in California. It delivered products to a customer's home within a 30-min period of their choosing. The company expanded to several other cities before it went bankrupt in 2001. Many of its failings were due to management as the business model was reasonable, and today there are several successful online fresh food delivery businesses. The management was inexperienced in the supermarket or grocery business, and the company spent excessively on infrastructure. It had been advised to build up an infrastructure to deliver groceries as quickly as possible, rather than developing partnerships with existing supermarkets. It built warehouses, purchased a fleet of delivery vehicles and top of the range computer infrastructure before running out of money.

Ernst Malmsten and others founded Boo.com in 1998, as an online fashion retailer that was based in the UK. The company spent over $135 million of shareholder funds in less than 3 years, before it went bankrupt in 2000. Its website was poorly designed for its target audience, and it went against many of the accepted usability conventions of the time. The website was designed in the days before broadband, with 56 K modems used by most customers. However, its design included the latest Java and Flash technologies, and it took most users several minutes to load the first page of the website. Further, the navigation of the website was inconsistent and changed as the user moved around the site.

Other reasons for failure included poor management and leadership, lack of direction, lack of communication between departments, spirally costs left unchecked and crippling payroll costs. Further, purchasers returned many products, and there was no postage charge applied for this service. The company went bankrupt in 2000, and an account of its formation and collapse is in the book, *Boo Hoo* (Malmsten and Portanger 2002). This book is a software development horror story, and the maturity of the software development practices employed may be judged from the fact that the developers were working without any source code control mechanism in place (a basic software engineering practice). The net effect was that despite extensive advertising by the company users were not inclined to use the site.

Pets.com was an online pet supply company founded in 1998 by Greg McLemore. It sold pet accessories and supplies, and it had a well-known advertisement as to '*why one should shop at an online pet store?*'. The answer to this question was: '*Because Pets Can't Drive!*'. Its famous mascot (the Pets.com dog sock puppet) was used in its marketing campaign. It launched its IPO in February 2000 just before the dot-com collapse.

Pets.com made investments in infrastructure such as warehousing and vehicles. It needed a critical mass of customers to break even and its management believed that it needed $300 million of revenue to achieve this. They expected that this would take a minimum of 4–5 years, and therefore there was a need to raise further capital. However, following the dot-com collapse, there was negative sentiment towards technology companies, and it was apparent that it would be unable to raise further capital. The management tried to sell the company without success, and it went into liquidation 9 months after its IPO.

Joseph Park and Yong Kang founded Kozmo.com in New York in 1998 as an online company that promised free 1-hour delivery of small consumer goods. It provided point-to-point delivery (usually on a bicycle) and did not charge a delivery fee. Its business model was deeply flawed, as it is expensive to offer point-to-point delivery of small goods within a 1-hour period without charging a fee. The company argued that they could make savings to offset the delivery costs, as they did not require retail space. It expanded into several cities in the United States, and raised about $280 million from investors. The company ceased trading in 2001.

13.7.2   Business Models

A business model converts a business or technology idea into a commercial reality, and it needs to be appropriate for the company and its intended operating market. A company with an excellent business idea but with a weak business model may fail, whereas a company with an average business idea but an excellent business model may be quite successful. Several of the business models in the dot-com era were deeply flawed, and the eventual collapse of many of these companies was predictable. Chesbrough and Rosenbloom (Chesbrough and Rosenbloom 2002) have identified six key components in a business model (Table 13.4):

13.7.3   Bubble and Burst

The initial public offering of Netscape in 1995 demonstrated the incredible value of the new Internet companies. Netscape had planned to issue the share price at $14, but it decided at the last minute to issue it at $28. The share price reached $75 later

Table 13.4 Characteristics of business models

Constituent	Description
Value proposition	This describes how the product or service is a solution to a customer problem
Market segment	This describes the customers that will be targeted (including market segments)
Value chain structure	This describes where the company fits into the value chain (Porter 1998)
Revenue generation and margins	This describes how revenue will be generated, including revenue streams from sales, support, etc.
Position in value network	This involves identifying competitors and other players that can assist in delivering added value to the customer
Competitive strategy	This describes how it will develop a competitive advantage to be successful

that day. This was followed by what became the dot-com bubble where there were many public offerings of Internet stock, and the value of these stocks reached astronomical levels. Reality returned to the stock market when it crashed in April 2000, and share values returned to more realistic levels.

Most of these Internet companies were losing substantial sums of money, and few expected to deliver profits in the short term. Financial instruments such as the balance sheet, profit and loss account, and price to earnings ratio are normally employed to estimate the value of a company. However, investment bankers argued that there was a new paradigm in stock market valuation for Internet companies. This paradigm suggested that the potential future earnings of technology companies be considered in determining their value, and this was used to justify the high prices of shares, as frenzied investors rushed to buy these over-priced and over-hyped stocks. Common sense seemed to play no role in decision-making. The dot-com bubble was characterised by:

- Irrational exuberance on the part of investors.
- Insatiable appetite for Internet stocks.
- Incredible greed from all parties involved.
- Following herd mentality.
- A lack of rationality and common sense by all concerned.
- Traditional method of company valuation not employed.
- Interest in making money rather than in building the business first.
- Questionable decisions by Federal Reserve Chairman (Alan Greenspan).
- Questionable analysis by investment firms.
- Investment banks had conflicts of interest and did not question the boom too closely.
- Market had left reality behind.

There were winners and losers in the boom and collapse. Some investors made a lot of money from the bubble, with others including pension funds and life assurance funds making significant losses. The investment banks typically earned 5–7% commission on each successful IPO, and it was not in their interest to question the boom too closely. Those who bought and disposed early obtained a good return, whereas those who kept their shares for too long suffered losses. The full extent of the boom can be seen in the rise and fall of the value of the Dow Jones and NASDAQ from 1995 through 2002.

The extraordinary rise of the Dow Jones (Fig. 13.2) from a level of 3800 in 1995 to 11,900 in 2000 represented a 200% increase over 5 years or approximately 26% annual growth (compound) during this period. The rise of the NASDAQ (Fig. 13.3) over this period is even more dramatic. It rose from a level of 751 in 1995 to 5000 in 2000 representing a 566% increase during the period. This is equivalent to a 46% compounded annual growth rate of the index.

The fall of the indices was equally as dramatic especially in the case of the NASDAQ. It peaked at 5000 in March 2000, and fell to 1200 (a 76% drop) by September 2002. It had become clear that Internet companies were rapidly going

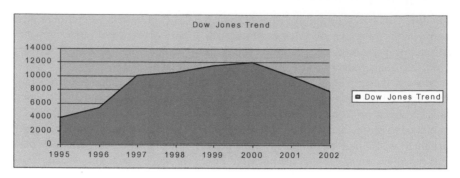

Fig. 13.2 Dow Jones (1995–2002)

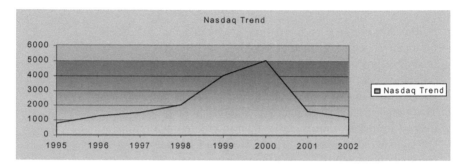

Fig. 13.3 NASDAQ (1995–2002)

through the cash raised at the IPOs, and analysts noted that a significant number would be out of cash by the end of 2000. Therefore, these companies would either go out of business or would need to go back to the market for further funding. This led to questioning of the hitherto relatively unquestioned business models of these Internet firms. Funding is easy to obtain when stock prices are rising at a rapid rate. However, when prices are static or falling, with negligible or negative business return to the investor, then funding dries up. The actions of the Federal Reserve in rising interest rates to prevent inflationary pressures also helped to correct the irrational exuberance of investors.

Some independent commentators had recognised the bubble but their comments and analysis had been largely ignored. These included 'The Financial Times' and the 'Economist' as well as some commentators in the investment banks. Investors rarely queried the upbeat analysis coming from Wall Street and seemed to believe that rising stock prices would be a permanent feature of the US stock markets. Greenspan had argued that it is difficult to predict a bubble until after the event, and that even if the bubble had been identified it could not have been corrected without causing a contraction. Instead, the responsibility of the Fed (according to Greenspan) was to mitigate the fallout when it occurs.

There have, of course, been other stock market bubbles throughout history. For example, in the 1800s there was a rush on railway stock in England leading to a bubble and eventual burst of railway stock prices in the 1840s. There was a devastating property bubble and collapse (2002–2009) in the Republic of Ireland. The failure of the Irish political class, the Irish Central bank and financial regulators, the Irish Banking sector in their irresponsible lending policies, and failures of the media in questioning the bubble are deeply disturbing. Its legacy remains and while the country has made a remarkable recovery, the failures of so many at senior level in the state remain deeply disturbing.

13.8 E-Commerce Security

The World Wide Web consists of unknown users and suppliers with unpredictable behaviour operating in unknown countries around the world. These users and websites may be friendly or hostile and the issue of trust arises:

- Is the other person who they claim to be?
- Can the other person be relied upon to deliver the goods on-payment?
- What legal remedies are there if the goods are not delivered?
- Can the other person be trusted not to inflict malicious damage?
- Is financial information kept confidential on the server?

Hostility may manifest itself in various acts of destruction. For example, malicious software may attempt to format the hard disk of the local machine, and if successful all local data will be deleted. Other malicious software may attempt to steal confidential data from the local machine including bank account or credit card details. The *denial-of-service attack* is when a website is overloaded by a malicious attack and where users are unable to access the website for an extended period.

The display of web pages on the local client machine may involve the downloading of programs from the server, and running the program on the client machine. Standard HTML allows the static presentation of a web page, whereas many web pages include active content (e.g. Java Applets or Active X). There is a danger that a Trojan horse[4] may be activated during the execution of active content.

Security threats may be from anywhere (e.g. client side, server side and transmission) in an e-commerce environment, and therefore a holistic approach is

---

[4]The origin of the term 'Trojan Horse' is from Homer's Illiad and concerns the Greek victory in the Trojan war. The Greek hero, Odysseus and others hid in a wooden horse while the other Greeks sailed away from Troy. This led the Trojans to believe that the Greeks had abandoned their attack and were returning to their homeland leading behind a farewell gift for the citizens of Troy. The Trojans brought the wooden horse into the city and later that night Odysseus and his companions opened the gates of Troy to the returning Greeks, leading to the mass slaughter of its citizens. Hence, the phrase '*Beware of Greeks bearing gifts*'. Troy was located at the mouth of the Dardanelles in Turkey.

required to protect the user. Internal and external security measures need to be considered, with internal security generally implemented with good processes and procedures and assigning appropriate access privileges.

It is essential that users have confidence in the security provided as otherwise they will be reluctant to pass credit card details over the web for purchases. Technologies such as secure socket layer (SSL) and secure HTTP (S-HTTP) help to ensure security.

13.9   Internet of Things

The Internet of Things refers to interconnected technology that is now an integral part of modern society, where computation and data communication are embedded in our environment. The Internet of Things is not a single technology as such, and instead it is a collection of devices, sensors and services that capture data to monitor and control the world around them. It means that information processing is now an integral part of people's lives.

The Internet of Things has been applied to several areas including our bodies (quantified self), our homes (smart homes) and public spaces (smart city). Wearable biometric sensors may be used to determine the calories burned during a period of exercise, as well as monitoring heart rate, breathing, skin temperature and perspiration. In theory, this helps individuals to control key parameters associated with their health.

The rise of the *smart home* is intended to deliver convenience to home occupiers, and these consist of connected devices that provide useful functionality. The various digital controls in a modern home may be used to control lighting, entertainment and security, as well as cooling and ventilation systems. The devices gather data about the environment as well as passing data back to the service provider. However, there are dangers with giving all this data about your life to a service provider, as it is essential that the privacy of an individual is protected.

The rise of the *smart city* is where the modern city collects data about its inhabitants and uses it to make more efficient use of energy, space and other resources. The data may be gathered through CCTV and other devices, and in the future the smart city will have knowledge of the habits and energy use of the citizens, allowing it to control resources more effectively.

There are several implicit assumptions with respect to smart cities, and it seems to be assumed that it is possible to know all aspects of the world perfectly with data, that the data will always be accurate and that the data will be easy to interpret. These assumptions are questionable.

That is, while the Internet of Things presents new possibilities, it is important to proceed cautiously and to use it sensible as an extra tool that may support decision-making rather than assuming that it provides all the answers. For further information on the Internet of Things, see the thought-provoking Guardian article (Greenfield 2017).

13.10   Internet of Money and Bitcoin

The idea of the Internet of Money is to build a financial environment that is suitable for the Internet world, and it moves away from the traditional centralised model where third-party banks record and manage all financial transactions. The new paradigm is a decentralised model via the Internet where buyers and sellers interact directly through *digital currencies* and decentralised ledgers. This decentralised model is termed the '*Internet of Money*', and *Bitcoin* aims to satisfy this model.

Digital currency is a type of currency that is available only in digital form, and it exhibits properties like the traditional physical currencies in that they may be used to buy goods and services. They include virtual currencies and cryptocurrencies.

The concept of digital cash was proposed by David Chaum in the early 1980s, and he formed DigiCash (an electronics cash company) in the early 1990s to commercialise his research (Chaum 1982). The goal of electronic cash (ecash) is to allow the user to be anonymous, and it allows users to spend in a manner that is untraceable to a bank or any other third party.

Chaum introduced the idea of blind signatures in his 1982 paper, which blinds the content of a message before it is signed. This means that the signer cannot determine the content of the message, but the resulting blind signature can be verified against the original unblinded message.

One of the earliest digital currencies was e-Gold (it was backed by Gold), and this centralised service appeared in the mid-1990s. It was later shut down by the US government, due to concerns over money laundering. Q coins emerged around 2005 and Bitcoin appeared in 2008. Bitcoin is the most widely used and accepted digital currency, and it is based on cryptographic algorithms (i.e. it is a cryptocurrency).

There are several types of digital currency including centralised systems (e.g. PayPal and eCash) which sell digital currency directly to the end user, mobile digital wallets for contactless payment transfer to facilitate easy payment (e.g. Google Wallet and Apple Pay make it easy to carry all your debit and credit cards on your smartphone) and decentralised system which employ cryptocurrencies and rely on cryptography (Bitcoin is the most well known of these). Finally, there are virtual currencies which are issued and controlled by its developers, and accepted by the members of a virtual community.

Bitcoin is a cryptocurrency and digital payment system, and it is the first decentralised digital currency. It was created by an unknown inventor(s) with the pseudonym Satoshi Nakamoto in 2008 (Nakamoto 2008), and it works without a central repository or single administrator. It is peer-to-peer with transactions taking place directly between users without the need for third-party intermediaries, and the transactions are verified by network nodes and recorded in a public distributed ledger termed a blockchain. The open-source software for Bitcoin was released by Nakamoto in early 2009, and the domain name bitcoin.org was registered in 2008.

The unit of account in the Bitcoin system is the bitcoin (BTC) with smaller amounts represented by millibitcoins (0.001 BTC), and the smallest amount is the satoshi (0.00000001 BTC).

13.11   Review Questions

1. Describe the development of the Internet.
2. Describe the development of the World Wide Web and its key constituents.
3. Describe the applications of the World Wide Web.
4. Describe the key constituents of an electronic commerce site.
5. Describe a successful dot-com company that you are familiar with. What has made the company successful?
6. Describe a dot-com failure that you are familiar with. What caused the company to fail?
7. Discuss the key components of a business model.
8. Discuss security in an e-commerce environment.

13.12   Summary

This chapter considered the evolution of the Internet from the early work on packet switching and ARPANET, to the subsequent development of the TCP/IP network protocol which is a transmission standard that deals with physical differences in host computers, routers and networks.

TCP/IP is designed to transfer data over networks which support different packet sizes and which may sometimes lose packets. TCP details how information is broken into packets and re-assembled on delivery, whereas IP is focused on sending the packet across the network.

The invention of the World Wide Web by Tim Berners-Lee was a revolutionary milestone in computing. It transformed the Internet from mainly academic use to commercial use, and it led to a global market of consumers and suppliers. Today, the World Wide Web is an integral part of peoples' lives.

The growth of the World Wide Web was exponential, and it led to the formation of many 'new economy' businesses. These new companies conducted business over the web as distinct from the traditional bricks and mortar companies. Some of these new companies were very successful (e.g. Amazon) and remain in business. Others were financial disasters due to poor business models, poor management and poor implementation of the new technology.

The dot-com bubble was characterised by many public offerings of Internet stock, where the value of these stocks reached astronomical levels. Reality returned to the stock market when the bubble burst and the market crashed in 2000. Finally, we discussed the Internet of Things and the Internet of Money.

Chapter 14
The Smartphone and Social Media

Key Topics

PDA
Smartphone
Facebook
Tweets
Twitter

14.1 Introduction

Smartphones arose as the outcome of the marriage of the existing mobile phone technology and PDA technology, and they contain advanced computing capabilities that are attractive to users. Today, the smartphone is ubiquitous, with most people in advanced countries owning one.

The introduction of the PDA by Apple and Palm played a role in the development of the smartphone, and its introduction facilitated a major growth of social networking. Users were now able to communicate news events or update their personal information in real time. Social networking sites such as Facebook and Twitter have transformed human communication.

Social media involves the use of computer technology for the creation and exchange of user-generated content. These web-based technologies allow users to discuss and modify user-created content, and it has led to major changes in communication between individuals, communities and organisations.

Facebook helps users to keep in touch with friends and family, and it allows them to share their opinions on what is happening around the world. Users may upload photos and videos, express opinions and ideas, and exchange messages. Facebook allows the user's community of friends to be actively kept up to date on important events that the user wishes to share.

Facebook has become an important communication channel for educated young people to discuss their aspirations for the future, as well as their grievances with society and the state. It has become an effective tool for protest and social revolution.

© Springer International Publishing AG, part of Springer Nature 2018
G. O'Regan, *World of Computing*,
https://doi.org/10.1007/978-3-319-75844-2_14

Twitter has become an effective way to communicate the latest news and its effectiveness as a communication tool increases as the number of a person's followers grows. It allows a person or organisation to determine what people are saying about it, including their positive or negative experiences. This allows direct interaction with the followers, and so it is a powerful way to engage the audience and to make people feel heard.

14.2   Evolution of the Smartphone

A smartphone is more than a mobile device for making and receiving calls, and it is essentially a touch-based computer on a phone, which comes with its own touch-screen keyboard, operating system, Internet access and third-party applications. It provides many other features such as a camera, maps, browser, email, calendar, alarm clock and games.

IBM (in a joint venture with BellSouth) introduced one of the earliest precursors of today's smartphones in 1993. This was the IBM Simon, and it included voice and data services. It acted as a mobile phone, a PDA and a fax machine, and it also included a touchscreen that could be used to dial numbers. It could send faxes and emails as well as making or receiving calls, and it included applications such as an address book, calendar and calculator. However, it was an expensive and large bulky device, and it was priced at $900.

John Sculley, the CEO of Apple, coined the term *'Personal Digital Assistant'*, and Apple introduced the first PDA, the Newton, in 1993. The Apple Newton included some nice features including limited handwriting recognition abilities. Xerox PARC had created a prototype PDA, the Dynabook, in the 1970s, but they did not commercialise it.

A PDA allows a large amount of data to be stored on a small handheld device. Palm introduced an early PDA device, the Palm Pilot 1000, in 1996, and this was used for mobile data. It played an important role in popularising the use of mobile data by business users. The Palm Pilot started the PDA industry, and it included 128 Kb of memory and 16 MHz of processing power. It had better handwriting recognition capabilities than the Newton and a graphical user interface (GUI).

The Nokia 9000 Communicator was released in 1996, and this phone combined the features of a PDA and a mobile phone. It included a physical QWERTY keyboard, and it provided features such as email, calendar, address book and calculator. However, it did not provide the ability to browse the web, and a colour display was introduced in the Nokia 9120 in 1998.

Qualcomm introduced its pdQ smartphone in 1999, and this phone combined a Palm PDA with Internet connectivity capabilities. Research in motion (RIM) released its first Blackberry devices in 1999, and these provided secure email communication into a single inbox. Samsung's first smartphone was the Samsung SPH-I300, which was released in 2001, and this Palm-powered smartphone is a distant ancestor of today's smartphones. Samsung introduced its SGH i607

Fig. 14.1 Apple iPhone 4

smartphone in 2006, and this Window's powered phone was inspired by Research in Motion's Blackberry phone.

Smartphone technology continued to evolve through the early 2000s, and Apple introduced its revolutionary *i*Phone in 2007. This Internet-based multimedia smartphone included a touchscreen, and features such as a video camera, email, web browsing, text messaging and voice. The *i*Phone had a 3.5 inch 480 × 320 touchscreen, a QWERTY touchscreen keyboard and 4 GB of storage. Apple developed its own operating system, *i*OS, for the *i*Phone.

Google introduced its open-source Android operating system in late 2007, and the first Android phone was introduced in late 2008. Android is now the dominant operating system for smartphones and tablets, with *i*OS used on Apple's products. The Samsung Instinct was released in 2008, but it was based on an operating system developed by Samsung from various Java components. Although its touchscreen operating system was not in the same league as Apple's *i*OS, it became a competitor to Apple's *i*Phone.

Apple's *i*Phone 4 (Fig. 14.1) was introduced in 2010, and this powerful smartphone has a 3.5 in. 960 × 640 screen and a 5-megapixel camera. The Samsung Galaxy S smartphone was launched in 2010, and this touchscreen-enabled Android smartphone became extremely popular. The Samsung Galaxy S series of smartphones have been very successful and have become a major competitor to Apple's *i*Phone.

Apple released the *i*Pad in 2010, which is a large screen tablet-like device that uses a touchscreen operating system. Samsung is a major competitor to Apple in the tablet market.

14.3 The Facebook Revolution

Facebook is the leading social networking site (SNS) in the world, and its mission is to make the world more open and connected. It helps users to keep in touch with friends and family, and it allows them to share their opinions on what is happening around the world. Users may upload photos and videos, express opinions and ideas, and exchange messages. Facebook is very popular with advertisers as it allows them to easily reach a large target audience.

Mark Zuckerberg (Fig. 14.2) founded the company in 2004 while he was a student studying psychology at Harvard University. Zuckerberg was interested in programming, and he had already developed several social networking websites for his fellow students including *Facemash* which could be used to rate the attractiveness of a person, and *Coursematch* which allowed students to view people taking their degree.

Zuckerberg launched '*The Facebook*' (thefacebook.com) at Harvard in February 2004, and over a thousand Harvard students had registered on the site within the first 24 h. Over half of the Harvard student population had a profile on Facebook within the first month. The membership of the site was initially restricted to students at Harvard, then to students at the other universities in Boston, and then to students at the other universities in the United States. Its membership was extended to international universities from 2005.

The use of Facebook was extended beyond universities to anyone with an email address from 2006, and the number of registered users began to increase exponentially. The number of registered users reached 100 million in 2008, 500 million in 2010, it exceeded 1 billion in 2012 and reached 2 billion in 2017. It is now one of the most popular websites in the world.

Facebook's business model is quite distinct from that of a traditional business in that it does not manufacture or sell any products. Instead, it earns its revenue mainly from advertisements, and its business model is based on advertisement revenue, with advertisements targeted to its over 2 billion users based on their specific interests. Facebook is essentially selling its users to advertisers (i.e. the users are the product). The users really do all the work, and Facebook collects data about them (e.g. age, gender, location, education, work history and interests) and classifies and categories

Fig. 14.2 Mark Zuckerberg

them, so that it may target advertisements that will potentially be of interest to them. This ensures that the advertisements are targeted to the right audience.

Social media have become important communication channels for educated young people to discuss their aspirations for the future, as well as their grievances with society and the state. The effectiveness of Facebook as a tool for protests and revolution is evident in the relatively short protests that culminated in the resignation of President Hosni Mubarak of Egypt in 2011.

Egypt has a young population with roughly 60% of the population under the age of 30, and the country has faced many challenges since independence such as improving education and literacy for its young population, as well as finding jobs for its citizens.

Facebook provided a platform for Egyptian youth to discuss issues such as unemployment, low wages, police brutality and corruption. Young Egyptians set up groups on Facebook to discuss specific issues (e.g. a group that aimed to provide solidarity with striking workers was set up). Further momentum for revolution followed the beating and killing of Khalid Mohammed Said, as photos of his disfigured body were posted over the Internet and went viral. An influential Facebook group called '*We are All Khalid Said*' was set up, and the killing provided a tangible focus for solidarity among young Egyptians.

The protests lasted for 18 days and it led to hundreds of thousands of young Egyptians taking to the streets and gathering in Tahrir Square in Cairo. They demanded an end to police brutality as well as the end of the 30-year reign of President Hosni Mubarak. The authorities reacted swiftly in closing the Internet in Egypt, but this act of censorship failed to stop the demonstrations and protests. Social media played an important role in mobilising protests, and in influencing the outcome of the revolution.

14.4 The Tweet

Twitter is a social communication tool that allows people to broadcast short messages. It is often described as the '*SMS of the Internet*', and Twitter is an online social media and micro-blogging site that allows its users to send and receive short 140-character messages called '*tweets*'. The restriction to 140 characters is to allow Twitter to be used on non-smartphone mobile devices.[1] Twitter has over 300 million active users, and it is one of the most visited websites in the world. Users may access Twitter through its website interface, a mobile device app or SMS.

Jack Dorsey (Fig. 14.3) and others founded the company in 2006. Dorsey introduced the idea of an individual using an SMS service to communicate with a small group while he was still an undergraduate student at New York University. The word '*twitter*' was the chosen name for this new service, and its definition as '*a short burst of information*' and '*chirps from birds*' was highly appropriate.

---

[1]Twitter plans to increase the character limit of the tweet to 280 characters.

Fig. 14.3 Jack Dorsey at the
2012 time 100 Gala

Twitter messages are often about friends telling one another about their day, what they are doing, where they are, whey they are thinking and doing, and Twitter has transformed the world of media, politics and business. It is possible to include links to web pages and other media as a tweet. News such as natural disasters, sports results and so on are often reported first by Twitter. The site has impacted political communication in a major way, as it allows politicians and their followers to debate and exchange political opinions. It allows celebrities to engage and stay in contact with their fans, and it provides a new way for businesses to advertise its brands to its target audience.

A Twitter user may select which other people that they wish to follow, and when you follow someone their tweets show up in a list known as your *Twitter stream*. Similarly, anyone that chooses to follow you will see your tweets in their stream.

A *hashtag* is an easy way to find all the tweets about a topic of interest, and it may be used even if you are not following the people who are tweeting. It also allows you to contribute to the topic that is of interest, and a hashtag consists of a short word or acronym preceded by the hash sign (#). Conferences, hot topics and so on often have a hashtag.

A word or topic that is tagged at a greater rate than other hashtags is said to be a *trending topic*, and a trending topic is often the result of an event that prompts people to discuss the topic. Trending may also result from the deliberate action of certain groups (e.g. in the entertainment industry) to raise the profile of a musician or celebrity and to market their work.

Twitter has evolved to become an effective way to communicate the latest news, and its effectiveness as a communication tool for an organisation increases as the number of its followers grows. An organisation may determine what people are saying about it, as well as their positive or negative experience in interacting with it. This allows the organisation to directly interact with its followers, which is a powerful way to engage with its audience and to make people feel heard. It allows the organisation to respond to any negative feedback and to deal with such feedback sensitively and appropriately.

The first version of Twitter was introduced in mid-2006, and it took the company some time to determine exactly what type of entity it was. There was nothing quite like it in existence, and initially it was considered a micro-blogging and social media site. Today, it is viewed as an information network rather than just a social media site.

Twitter has experienced rapid growth from 400,000 tweets posted per quarter in 2007, to 100 million per quarter in 2008, to 65 million tweets per day from 2010, to 140 million tweets per day in 2011 and to 500 million tweets per day in 2016. Twitter's usage spikes during important events such as major sporting events, natural disasters, the death of a celebrity and so on. For such events, there may be over 100,000 tweets per second.

Twitter's main source of revenue is advertisements through '*promoted tweets*' that appear in a user's timeline (Twitter stream). The first promoted tweets appeared from late 2011, and the use of a tweet for advertisement was ingenious. It helped to make the advertisement feel like part of Twitter, and it meant that an advertisement could go anywhere that a tweet could go. Advertisers are only charged when the user follows the links or re-tweets the original advertisements. Further, the use of tweets for advertisement meant that the transition to mobile was easy, and today about 80% of Twitter use is on mobile devices.

Twitter has recently embarked on a strategy that goes beyond these advertisements to sell products directly (including to people who do not use Twitter). Twitter also earns revenue from a data licensing arrangement where it sells its information to companies who use this information to analyse consumer trends. Twitter analyses what users tweet to understand their intent. For more detailed information on Twitter, see (Schaefer 2014).

14.5 Social Media and Fake News

Fake news is the systematic spreading of misleading or false information in traditional print or online social media, with the intention of misleading or damaging another person or institution. It can negatively affect individuals in a country and lead to violence or hate against minority ethnic groups. The popularity of social media sites such as Facebook have contributed to the spread of fake news, and this new phenomenon poses threats to twenty-first-century democracy. Fake news may be spread by individuals, organisations and hostile states, and it consists of news that has no basis in fact, but which is presented as being factually correct.

Fake news in the form of propaganda has been around for centuries, where such news is generally published for political reasons. Military leaders have often embellished their bravery and result in battle throughout history (e.g. Ramses II's description of the Battle of Kadesh in the thirteenth century B.C. paints a very positive but factually inaccurate account of the battle).

Following the invention of the printing press in the fifteenth century, news publications became popular, and over time fake news stories appeared in the print media. Fake news played an important role in propaganda during the first and second world wars, with radio broadcasts and printed material used to persuade the public at home as well as discouraging enemy troops. Today, modern society is highly dependent on accurate information in the print, radio, television and online media. The effectiveness of fake news increases when the stories spread widely (as often occurs in social media), and where users interact with and rely on these stories rather than on traditional news media.

Fake news played a key role in the 2016 presidential election in the United States, which led to the election of Donald Trump. Most of the fake election news in the last 3 months of the campaign were anti-Clinton, but it is very difficult to determine the extent to which this influenced the outcome of the election. Trump and his supporters seem to use the word 'fake news' to refer to the mainstream media that is opposed to him and his policies.

It is important when considering the accuracy of an article to consider the source of the news (e.g. is it written by a reputable news organisation such as the BBC or Reuters?), as well as considering the authenticity of its authors and the supporting sources. Fake news is a deeply disturbing Internet trend that needs to be resolved if technology is to serve humanity. Modern technology has provided many benefits to modern society, but it needs to be managed effectively.

Fake news is a dangerous trend in society, as false news can spread easily due to the speed and accessibility of modern technology. It allows individuals to be misled and negatively influenced. Online social media sites such as Facebook and Twitter have a responsibility to develop appropriate solutions to address this serious problem.

14.6   Review Questions

1. What is a PDA?
2. What is a smartphone?
3. What is social media? Explain how sites such as Facebook and Twitter have transformed human communication.
4. Explain how a company may use social media to market new products to its customers.
5. Explain how social media has been used as a tool for protest and revolution.
6. Why has Twitter been described as the SMS of the Internet?
7. Explain how social media has facilitated the spread of fake news.

14.7 Summary

A smartphone is essentially touch-based computer on a phone, which comes with its own keyboard, operating system, Internet access and third-party applications. It provides many other attractive features such as a camera, maps, calendar, alarm clock and games. It arose from the marriage of mobile phone technology and PDA technology.

The smartphone has facilitated a major growth of social networking, as users are now able to communicate news or update their personal information in real time. Social media involves the use of computer technology that allows the creation and exchange of user-generated content. It has led to major changes in communication between individuals, communities and organisations. Social networking sites such as Facebook and Twitter have transformed human communication.

Facebook helps users to keep in touch with friends and family, and it allows them to share their opinions on what is happening around the world. Users may upload photos and videos, express opinions and ideas, and exchange messages. It has become an important communication channel for young people to discuss their aspirations for the future, and it has also become an effective tool for mobilising protests and social revolution.

Twitter has become an effective way to communicate the latest news, and its effectiveness as a communication tool increases as the number of its followers grows. It allows a person or organisation to determine what people are saying about it, as well as their positive or negative experiences.

Chapter 15
Legal Aspects of Computing

Key Topics

Intellectual property
Patents, Copyright and trademarks
Computer crime
Hackers and privacy
Freedom of speech and censorship
Cyberextortion
Software licenses
Bespoke software
Ecommerce law
Plagiarism/censorship

15.1 Introduction

Legal aspects of computing are concerned with the application of the legal system to the computing field. This chapter explores several legal aspects of digital information and software, as well as legal aspects of the Internet.

We discuss intellectual property law including patents, copyright, trademarks and trade secrets. Patents provide legal protection for intellectual ideas; copyright law protects the expression of an idea; and trademarks provide legal protection of names or symbols.

We discuss the problem of hacking where a hacker is a person who uses his computer skills to gain unauthorised access to a computer system. We distinguish between ethical white hat hackers and malicious black hat hackers. We discuss computer crime including the unauthorised access of computer resources, the theft of personal information, cyberextortion and denial-of-service attacks.

A software license is a legal agreement between the copyright owner and the licensee, which governs the use or distribution of software to the user. The two most common categories of software licenses that may be granted under copyright law are those for proprietary software and those for free open-source software.

© Springer International Publishing AG, part of Springer Nature 2018
G. O'Regan, *World of Computing*,
https://doi.org/10.1007/978-3-319-75844-2_15

We discuss the legal aspects of bespoke software development, and a legal contract is prepared between the supplier and the customer. This will generally include a statement of work that stipulates the deliverables to be produced, and it may also include a service-level agreement and an Escrow agreement.

We discuss the nature of electronic commerce including transactions to place an order, the acknowledgement of the order, the acceptance of the order and order fulfilment.

15.2 Intellectual Property

Intellectual property law deals with the rules that apply in protecting inventions, designs and artistic work, and in enforcing such rights. Intangible assets such as software designs or inventions may be protected in a similar way to the protection of private property, and the inventor is generally granted exclusive rights to the invention for a defined period. This provides the inventor the incentive to develop creative works that may benefit society, as it allows the owner of the invention to profit from their work without fear of misappropriation by others.

The main forms of intellectual property are patents, copyright and trademarks. Patents give inventors exclusive rights to their invention for a specified period (possibly up to 20 years), or to profit from the invention by transferring the right to another party. A *patent* protects innovative ideas and concepts, and the invention itself must be novel and more than an obvious next step from existing technology. The patent needs to be filed at the Patent Office, and the patent gives the inventor protection against patent infringement in a specific country or region of the world.

A *copyright* applies to original writing, music, motion pictures and other original intellectual and artistic expressions. It does not protect the underlying idea as such, and what is protected is the expression of the idea. Copyrights are exclusive rights to making copies of the expression, where the ways of expressing ideas is copyrightable. Computer software source code is protected by copyright law. The term *'fair use'* refers to the permitted limited use of copyrightable material without acquiring permission from the copyright owner.

A *trademark* protects names or symbols that are used to identify goods or services, and their purpose is to avoid confusion and to help customers to distinguish one brand from another.

A *trade secret* is information that provides competitive advantage over others, and it is of value only if it is kept secret. It applies in the computer sector where programs may use algorithms that are unknown to others.[1]

----

[1]It is not illegal to use reverse engineering to try to discover the trade secret.

15.2.1   Patent Law

Patents are a part of intellectual property and they protect innovative ideas and concepts. A patent gives the inventor protection against infringement, and the inventor needs to file the patent at the Patent Office. The patent needs to be precisely described and to be successfully accepted the idea or invention must be novel (and more than an obvious next step from the existing technology). There needs to be a good business case for the patent (i.e. the idea must be such that competitors will need to use the invention and are unable to bypass the invention), and once the patent has been described at the right level of detail by the inventor, there will be a business decision on whether to file the patent at the Patent Office or not.[2]

The prosecution of the patent at the Patent Office will be done by a patent attorney, and it will require a detailed search to ensure that there is no existing *prior art* that would invalidate the patent application. Finally, the Patent Office grants (or rejects) the patent, and the inventor may then earn a royalty fee from the invention for a defined period (based upon its use).

A patent should have an informative title as well as a concise summary of the invention. It needs to provide a description of the current state-of-the-art as well as a technical description of the invention. It needs to highlight the applications and advantages of invention, and drawings should be included. It needs to employ clear wording and a glossary may be required. It is essential that the invention is *novel* and more than an obvious next step, and more than a transpose of existing technology. It needs a good business case, and it is desirable that competitors are unable to bypass the invention (as otherwise competitors will be able to avoid the payment of a license fee for its use).

The status of a patent application may be *unassessed* (if it has not been subject to a business review), *dropped* (the business review decides that it should not proceed any further), *filed* (an application has been made to the patent office), *prosecuted and granted* (the patent office has awarded the patent) and *defensive publication* (it has been decided not to file an application at the patent office, but to publish an article on the invention placing the invention in the public domain, thereby preventing a competitor from lodging a patent application).

15.2.2   Copyright Law

Copyrights apply to original writing and to original intellectual and artistic expressions, and it protects the expression of the idea rather than facts or the idea itself. Copyright law protects literary, musical and artistic works such as poetry,

---

[2]The decision may be to put the invention in the public domain with a defensive publication thereby preventing competitors from filing a patent for the invention.

songs, movies and computer software. It provides exclusive rights to making copies of the expression (subject to copyright law and fair use), where the ways of expressing ideas is copyrightable.

A copyright gives the copyright owner rights to exclude others from using or copying the finished work, and most copyrights are generally valid for the creator's lifetime plus 70 years (the exact period depends on the jurisdiction as copyright laws vary between countries).

The term *'fair use'* refers to the permitted limited use of copyrighted material without acquiring permission from the copyright owner. There are several factors that need to be considered before deciding whether fair use may be applied such as the purpose of use (e.g. non-profit educational use), the amount used (e.g. it is generally valid to use a small portion of the work for criticism or for education purposes), the amount used as a proportion of the whole of the copyrighted work and the effect of use on the market or value of the copyrighted work. The defendant bears the burden of proving fair use in any litigation on copyright infringement.

Computer software source code was granted protection by copyright law from the mid-1970s, which means that the reproduction of the computer software created by software developers and software companies is protected. The copyright grants the author the right to exclude others from making copies, and the owners of the copies have the right to make additional copies (for archival purposes) without the authorisation of the copyright owner. Further, owners of copies have the right to sell their copies.

This has led the software sector to move towards licensing their software rather than selling it. There is some software code that is freely available, and this includes software created by the free software movement (which began in the mid-1980s), the open-source initiative (which began in the late 1990s as a move that wished to highlight the benefits of freely available source code) or software that is in the public domain and that is therefore not subject to copyright. Open-source software (OSS) is software that is freely available under an open-source license to study, change and distribute to anyone for any purpose.

15.2.3 Trademarks

Trademarks protect names or symbols that are used to identify goods or services, and help customers to distinguish one brand from another. Trademark rights come from actual use, and a trademark does not expire after a fixed period provided it continues to be used. Brand names, slogans and logos are examples.

The registration of a trademark is not mandatory as rights to a mark may be granted based on its use. A registered trademark is indicated by$^{®}$, whereas an unregistered trademark is indicated by TM for goods and SM for services.

15.3   Hacking and Computer Security

A *hacker* is a person who uses his (or her) computer skills to gain unauthorised access to computer files or networks. A hacker may enjoy experimenting with computer technology (the original meaning of the term), but some hackers enjoy breaking into systems and causing damage (the modern meaning of the word). Ethical (*white hat*) hackers are former hackers who play an important role in the security industry in testing network security, and in helping to create secure products and services. Malicious (*black hat*) hackers (also called *crackers*) are generally motivated by personal gain, and they exploit security and system vulnerabilities to steal, exploit or sell data.

Many computer systems in use today have vulnerabilities that may be exploited by a determined hacker to gain unauthorised entry to the system, and access to unauthorised information. It is vital that best practice in software and system engineering is employed to develop safe and secure systems, and that known vulnerabilities in system security are addressed promptly by updates to the system software. Further, it is essential to educate staff on security, and to define (and follow) the appropriate procedures to prevent security breaches.

The early hackers were mainly young students (without malicious intent) who were exploring the university computer systems (such as the students at Massachusetts Institute of Technology in the late 1950s who were interested in exploring the IBM 704 computer), and they would enter areas of the system without authorisation and gain access to privileged resources. They were motivated by knowledge and wished to have a deeper understanding of the systems that they had access to. The idea of a hacker ethic was formulated in a book by Steven Levy in the mid-1980s (Levy 1984), and he outlined several key ethical principles including free access to computers and information and improvement to quality of life. Specifically, he defined six key tenets:

– Access to computers should be unlimited and total.
– All information should be free.
– Mistrust authority.
– Hackers should be judged by their hacking and not by bogus criteria such as race and religion.
– Art and beauty can be created on a computer.
– Computers can change your life for the better.

The *free software movement* arose in the early 1980s from followers of the hacker ethic, with Richard Stallman (its founder) often referred to as 'the last true hacker' (O'Regan 2015). Today, ethical hackers need to obtain permission prior to acting, as their actions may potentially cause major disruption to an organisation. Responsible (white hat) hackers can provide useful information on security vulnerabilities and may assist in improving computer security.

The security of the system refers to its ability to protect itself from accidental or deliberate external attacks, which are common today since most computers are networked and connected to the Internet. There are various security threats in any networked system including threats to the confidentiality and integrity of the system and its data, and threats to the availability of the system.

Therefore, controls are required to enhance security and to ensure that attacks are unsuccessful. Encryption is one way to reduce system vulnerability, as encrypted data is unreadable to the attacker. There may be controls that detect and repel attacks, and these controls are used to monitor the system and to take appropriate action to shut down parts of the system or restrict access in the event of an attack. There may be controls that limit exposure (e.g. insurance policies and automated backup strategies) that allow recovery from the problems introduced.

The introduction of the Internet in the early 1990s has transformed the world of computing, and it later led to an explosive growth in attacks on computers and systems, as hackers and malicious software sought to exploit known security vulnerabilities. It is therefore essential to develop secure systems that can deal with and recover from such external attacks.

Hackers will often attempt to steal confidential data and to disrupt the services being offered by a system. Security engineering is concerned with the development of systems that can prevent such malicious attacks and recover from them. It has become an important part of software and system engineering, and software developers need to be aware of the threats facing a system and develop solutions to manage them.

Hackers may probe parts of the system for weaknesses, and system vulnerabilities may lead to attackers gaining unauthorised access to the system. There is a need to conduct a risk assessment of the security threats facing a system early in the software development process, and this will lead to several security requirements for the system.

The system needs to be designed for security, as it is difficult to add security after the system has been implemented. Security loopholes may be introduced in the development of the system, and so care needs to be taken to prevent these as well as preventing hackers from exploiting security vulnerabilities.

The choice of architecture and how the system is organised is fundamental to the security of the system, and different types of systems will require different technical solutions to provide an acceptable level of security to its users. The following guidelines for designing secure systems are described in (Sommerville 2010).

− Security decisions should be based on the security policy.
− A security-critical system should fail securely.
− A secure system should be designed for recoverability.
− A balance is needed between security and usability.
− A single point of failure should be avoided.
− A log of user actions should be maintained.
− Redundancy and diversity should be employed.
− Organisation of information in system into compartments.

The unauthorised access to the system and the theft of confidential data and disruption of its services is unlawful, and we discuss computer crime in the next section.

15.4 Computer Crime

It is common in the major urban areas to encounter dangers in some streets or neighbourhoods, and such dangers need to be managed. Similarly, the Internet has dangers with hackers, scammers and web predators lurking in the shadows. A hacker may be accessing a computer resource without authorisation with the intention of committing an unlawful act. The hacker's activities may be limited to *eavesdropping* (listening to a conversation), or it may be an active *man-in-the-middle* attack, where the hacker may possibly alter the conversation between two parties.

One of the earliest Internet attacks was back in 1988 when a graduate student from Carnegie Mellon University released a program on the Internet (an Internet Worm) that exploited security vulnerability in the mail software to automatically replicate itself locally and on remote machines. It affected lots of machines and effectively shut down the Internet for 1–2 days.

Today, more and more individuals and companies are online, and networking systems and computers have become quite complex. There has been a major growth in attacks on businesses and individuals, and so it is essential to consider computer and network security. The Internet was developed based on trust with security features added as a response to different types of attacks.

There are several threats associated with network connectivity such as *unauthorised access* (a break-in by an unauthorised person), *disclosure of sensitive information* to people who should not have access to the information and *denial of service* (DoS), where there is a degradation of service that makes it impossible to access the website and perform productive work.

There may be attacks that lead to defacement of the websites, bank fraud, stealing of credit card numbers, hoax (scam) letters, phishing emails that appear to come from legitimate parties but contain links to a site that is different from the one that the user expects to go to, intercepting of packets and password sniffing. *Phishing* is an attempt to obtain sensitive information such as usernames, passwords and credit card details with the intention of committing fraud.

A computer *virus* is a self-replicating computer program that is installed on the user's computer without consent. It is a malicious software program that when executed replicates itself and infects other computer programs by modifying them. A virus often performs some type of harmful activity on the infected computers such as accessing private information, spamming email contacts or corrupting data. It is not a crime per se to write a computer virus or malicious software. However, if that software or other malware spreads to other computers, then it could be considered a crime.

Cyberextortion is a crime that involves an attack or threat of an attack, accompanied by a demand for money to stop the attack. They are often initiated through malware in an email attachment. These may include denial-of-service attacks or *ransomware* attacks that encrypt the victim's data. The victim is then offered the private key to resolve the encryption in return for payment. Companies need to manage the risks associated with cyberextortion and to ensure that end users are properly educated on malware and phishing.

Another form of computer crime is Internet fraud where one party is intent on deceiving another. Among these are hoax email scams, which are designed to deceive and fraud the email recipient. These may include the *Nigeria 419* scams, where the email recipient is offered a share of a large amount of money trapped in their country, if the recipient will help in getting the money out of the country. The recipient may be asked for their bank account details to help them to transfer the money (this information will later be used by them to steal funds), or the request may be to pay fees or taxes to release payment with further fees requested. Of course, the money will never arrive (*if an email looks like it really is too good to be true then it has a high probability of being a scam*).

15.5   Software Licensing

A software license is a legal agreement between the copyright owner and the licensee, which governs the use or distribution of software to the user (licensee). Computer software code is protected under copyright law in most countries, and a typical software license grants the user permission to make one or more copies of the software, where the copyright owner retains exclusive rights to the software under copyright law.

The two most common categories of software licenses that may be granted under copyright law are those for *proprietary software* and those for *free open-source software* (FOSS). The rights granted to the licensee are quite different for each of these categories, where the user has the right to copy, modify and distribute (under the same license) software that has been supplied under an open-source license, whereas proprietary software typically does not grant these rights to the user.

The *licensing of proprietary software* typically gives the owner of a copy of the software the right to use it (including the rights to make copies for archival purposes). The software may be accompanied with an end-user license agreement (EULA) that may place further restrictions on the rights of the user. There may be restrictions on the ownership of the copies made, and on the number of installations allowed under the term of the distribution. The ownership of the copy of the software often remains with the copyright owner, and the end user must accept the license agreement to use the software.

The most common licensing model is per single user, and the customer may purchase a certain number of licenses over a fixed period. Another model employed is the license per server model (for a site license), or a license per dongle model, which allows the owner of the dongle use the software on any computer. A license may be perpetual (it lasts forever) or it may be for a fixed period (typically 1 year).

The software license often includes maintenance for a period (typically 1 year), and the maintenance agreement generally includes updates to the software during that time and it may also cover a limited amount of technical support. The two parties may sign a service-level agreement (SLA), which stipulates the service that will be provided by the service provider. This will generally include timelines for the resolution of serious problems, as well as financial penalties that will be applicable where the customer service performance does not meet the levels defined in the SLA.

Free and open-source licenses are often divided into two categories depending on the rights to be granted in distribution of the modified software. The first category aims to give users unlimited freedom to use, study and modify the software, and if the user adheres to the terms of an open-source license such as GNU or General Public License (GPL), the freedom to distribute the software and any changes made to it. The second category of open-source licenses gives the user permission to use, study and modify the software, but not the right to distribute it freely under an open-source license (it could be distributed as part of a proprietary software license).

15.5.1   *Software Licensing and Failure*

Software license agreements generally include limited warranties on the quality of the licensed software, and they often provide limited remedies to the customer when the software is defective. The software vendor typically promises that the software will conform to the software documentation for a specified period (the warranty period), and the software warranty generally excludes problems that are not caused by the software or are beyond the software vendor's control.

The customers are generally provided with limited remedies in the case of defective software (e.g. the replacement of the software with a corrected version, or termination of the user's right to use the defective software and a refund of the license fee). The payment of compensation for loss or damage is generally excluded in the software licensing agreement.

Software licensing agreements are generally accompanied by a comprehensive disclaimer that protects the software vendor from any liability (however remote) that might result from the use of the software. It may include statements such as '*the software is provided "as is", and that the customers use the software at their own risk*'.

A limited warranty and disclaimer limits the customer's rights and remedies if the licensed software is defective, and so the customer may need to consider how best to manage the associated risks.

15.6 Bespoke Software Development

Bespoke software (or custom software) is software that is developed for a specific customer or organisation, and it needs to satisfy the defined customer requirements. The organisation will need to be rigorous in its selection of the appropriate supplier, as it is essential that the supplier selected has the capability of delivering high-quality and reliable software on time and on budget.

This means that the capability of the supplier is clearly understood and the associated risks are known prior to selection. The selection is based on objective criteria such as cost, the approach, the ability of the supplier to deliver the required solution, the supplier capability, and while cost is an important criterion, it is just one among several other important factors.

Once the selection of the supplier is finalised a legal agreement is drawn up between the contractor and supplier, which states the terms and condition of the contract, as well as the statement of work. The *statement of work* (SOW) details the work to be carried out, the deliverables to be produced, when they will be produced, the personnel involved their roles and responsibilities, any training to be provided and the standards to be followed. The agreement will need to be signed by both parties and may (depending on the type of agreement) include:

- Legal contract,
- Statement of work,
- Implementation plan,
- Training plan,
- User guides and manuals,
- Customer support to be provided,
- Service-level agreement,
- Escrow agreement and
- Warranty period.

A *service-level agreement* (SLA) is an agreement between the customer and service provider which specifies the service that the customer will receive as well as the response time to customer issues and problems. It will also detail the penalties should the service performance fall below the defined levels.

An *Escrow agreement* is an agreement made between two parties where an independent trusted third party acts as an intermediary between both parties. The intermediary receives money from one party and sends it to the other party when contractual obligations are satisfied. Under an Escrow agreement, the trusted third party may also hold documents and source code.

15.7   Dark Side of the Internet

The Internet has a dark and secret side where harmless or sinister activities may be conducted. These include online services such as online pornography and adult chat rooms, escort sites and so on. There are more sinister sites where a consenting adult unintentionally downloads malicious software from an adult chat room which infects the computer and allows someone to hack into the machine's camera, and the adult is captured on camera performing compromising acts and is then contacted by the gang or fraudster with demands for a payout (sextortion) to prevent the images and video being made public.

Other unsavoury activities include revenge porn where one of the parties to the relationship releases private images/videos of their former partner as an act of revenge at the end of the relationship. Sexting is where the sender sends privates images (of himself or herself) to the recipient, and the recipient makes the images available publicly (betrayal of trust and the naivety of the sender).

Other distasteful activities include cyberbullying where a child or young adult is bullied online by his or her peers, and sometimes there are devastating consequences. The Internet is a great resource but care is required to avoid being a victim of its dark side, and this requires education on its dangers as well as on its many positive aspects.

15.8   Ecommerce and the Law

The invention of the World Wide Web led to a revolution in business with commerce conducted online over the Internet, and today ecommerce sites are ubiquitous with business marketing and selling their products to customers around the world. Ecommerce uses several technologies such as the Internet and World Wide Web, electronic funds transfer, supply chain management and inventory management systems.

The ecommerce website needs to be carefully designed so that users can easily navigate its catalogue of products and make an informed decision on which produces to purchase. Further, the website needs to provide information about the business, its address and contact details, its products and their prices, the shipping costs and so on.

The user may select several products/services to purchase and may place an electronic order (the website will include an order/buy button). An acknowledgement email is sent shortly after placing the order (this is confirming that an order has been received but it is not confirming acceptance of the order). A separate email is sent confirming acceptance of the order, and this is confirmation of the electronic transaction between both parties (*a contract now exists between both parties*). The terms and conditions of purchase are specified either on the website or included in the email confirming the acceptance of the order. This will include information on the delivery period as well as the consumer's cancellation rights (during the cooling off period). Further, the terms of purchase must be fair and reasonable and written clearly, with unfair terms legally unenforceable.

The *law of tort* refers to a civil wrong where one party (the *defendant*) is held accountable for their actions (by the *plaintiff*). There are several actions that the defendant could be held accountable, e.g. negligence, trespass, misstatement, product liability, defamation and so on. For example, the defendant may be accused of negligence and a breach of his duty of care, where damage that was reasonably foreseeable was caused by negligence.

The Internet is global with business conducted internationally over the World Wide Web. However, a business is subject to the ecommerce laws of the country in which it is doing business, i.e. the law is National, and so a business needs to be familiar with and follow the specifics of the ecommerce law in the particular jurisdiction in which it is doing business. Ecommerce law is not radically different from standard commerce law, and in general, it follows the same basic principles. For example, false advertising and copyright infringement are not allowed, and if an item cannot be sold in a physical shop in a given country, then it cannot legally be sold online in that country.

Privacy is one area where ecommerce law differs from commerce law, since an online business collects a lot of information about the customer (financial and non-financial). It is essential that the online business has an appropriate privacy policy and that it protects the privacy of the customer's information. *Data protection* law refers to laws that define the ways in which information about living people may be legally used, with the goal of protecting people from the misuse or abuse of their personal information.

15.9   Free Speech and Censorship

Free speech and censorship are the opposites of one another, with censorship viewed as the suppression of free speech. Censorship is concerned with suppressing or removing anything deemed objectionable (e.g. obscene or indecent material). For example, television networks often bleep out swear words that are potentially offensive to their audience. Several countries around the world control access to specific websites, and thereby restrict the information that their citizens may receive. For example, the Great Firewall of China (GFW) regulates the use of the Internet in China, and it blocks access to selected foreign websites (e.g. Google and Facebook). State censorship is often concerned with controlling the population and preventing free expression that could lead to an uprising against the state.

Freedom of speech (or expression) is concerned with the right to express one's opinions and ideas without fear of censorship or government sanction. It is recognised as a right under Article 19 of the UN Declaration of Human rights, but it is a right that is subject to special duties and responsibilities (i.e. *freedom of speech is not an absolute right*, and is subject to limitations such as libel, slander, obscenity, defamation, hate speech, public order and incitement to violence). Freedom of speech is a key tenet of western democracies, but it also places responsibilities on the citizens.

Social media sites such as Facebook are testing the legal boundaries of free speech, and the question is how far a person's public speech may inflame its audience before it may be restrained. The use of Facebook plays an important role as a tool for social protest and revolution (see Chap. 14), and it played an important role in the 2011 revolution that led to the overthrow of President Hosni Mubarak of Egypt.

15.10   Computer Privacy in the Workplace

The right of an employee to privacy in the workplace has become more controversial in the digital age. Employers now have technology to monitor the use of computer resources in the workplace and may monitor communications such as the Internet and electronic mail. Employees may naturally feel that such monitoring is a violation of their privacy, but employee activities when using an employer's computer systems are generally not protected by privacy laws.

That is, emails are company property, and employers generally have the right to monitor and view such emails to check for productivity, illegal use and so on. Further, emails may be used as evidence in cases of proof of employee misconduct or wrongdoing.

Further, employers have the right to track websites visited by their employees to ensure that an employee is not spending an excessive amount of time at a specific site. Employers have the right to block or limit the time that an employee may spend online.

Employee has rights to privacy in the workplace but these need to be balanced against the employer's right to monitor its business operations. That is, while it is reasonable for an employer to monitor email and Internet use to ensure that employees do not abuse it and that the business is operating effectively, an employee has reasonable rights of privacy if computer resources are used appropriately. And so, it is about balancing rights given that on the one hand, the employer is paying the employee's salary and has a reasonable expectation that the employee does not abuse email and the Internet, and on the other hand the employee has reasonable expectations of privacy provided that the computer resources are used appropriately and not abused.

15.11   Review Questions

1. What is intellectual property law?
2. Explain the difference between a patent, copyright and trademark.
3. What is computer crime?
4. Explain how software is licensed to users.
5. Explain cyberextortion.
6. Explain the legal aspects of bespoke software development.
7. Explain the difference between ethical and malicious hackers.
8. Discuss ecommerce law and standard commerce law.

15.12   Summary

Legal aspects of computing are concerned with the legal aspects of digital information and software, as well as the legal aspects of the Internet. It deals with Intellectual property law including patents, copyright, trademarks and trade secrets.

Computer crime includes the unauthorised access to computer resources, the theft of personal information and denial-of-service attacks. A hacker uses his computer skills to gain unauthorised access to a computer system. Computer crime also includes Internet fraud, cyberextortion and viruses.

A software license is a legal agreement between the copyright owner and the licensee, which governs the use or distribution of software to the user (licensee). Computer software code is protected under copyright law, and the license grants the user permission to make one or more copies of the software. Software license agreements generally include limited warranties on the quality of the software and provide limited remedies to the customer when the software is defective.

Bespoke software (or custom software) is software that is developed for a specific customer or organisation, and needs to satisfy specific customer requirements. An appropriate supplier is selected, and a legal agreement is drawn up, which states the terms and condition of the contract, as well as the statement of work.

Ecommerce law is concerned with laws to regulate online electronic transactions and to protect the rights of consumers. Freedom of speech and censorship are opposite sides of the same coin, with censorship concerned with the suppression of free speech.

Chapter 16
Ethics and Professional Responsibility

Key Topics

Ethics
Parnas on Professional Responsibility
ACM Code of Ethics and Professional Practice
BCS Code of Conduct
Licensing of Software Engineers
Professional Conduct

16.1 Introduction

Ethics is a practical branch of philosophy that deals with moral questions such as what is right or wrong, and how a person should behave in a given situation in a complex world. Ethics explores what actions are right or wrong within a specific context or within a certain society, and seeks to find satisfactory answers to moral questions. The origin of the word 'ethics' is from the Greek word ἠθικός, which means habit or custom.

There are various schools of ethics such as the *relativist* position (as defined by Protagoras), which argues that each person decides on what is right or wrong for them; *cultural relativism* argues that the particular society determines what is right or wrong based upon its cultural values; *deontological ethics* (as defined by Kant) argues that there are moral laws to guide people in deciding what is right or wrong; and *utilitarianism* argues that an action is right if its overall effect is to produce more happiness than unhappiness in society.

Professional ethics are a code of conduct that governs how members of a profession deal with each other and with third parties. A professional code of ethics expresses ideals of human behaviour, and it defines the fundamental principles of the organisation and is an indication of its professionalism. Several organisations such as the Association Computing Machinery (ACM) and British Computer Society (BCS) have developed a code of conduct for their members, and violations of the code by members are taken seriously and are subject to investigations and disciplinary procedures.

© Springer International Publishing AG, part of Springer Nature 2018   281
G. O'Regan, *World of Computing*,
https://doi.org/10.1007/978-3-319-75844-2_16

Business ethics define the core values of the business and are used to guide employee behaviour. Should an employee accept gifts from a supplier to a company as this could lead to a conflict of interest? A company may face ethical questions on the use of technology. For example, should the use of a new technology be restricted because people can use it for illegal or harmful actions as well as beneficial ones?

Consider mobile phone technology, which has transformed communication between people, and thus is highly beneficial to society. What about mobile phones with cameras? On the one hand, they provide useful functionality in combining a phone and a camera. On the other hand, they may be employed to take indiscreet photos without permission of others, which may then be placed on inappropriate sites. In other words, how can citizens be protected from inappropriate use of such technology?

16.2 Business Ethics

Business ethics (also called corporate ethics) is concerned with ethical principles and moral problems that arise in a business environment. They refer to the core principles and values of the organisation, and apply throughout the organisation. They guide individual employees in carrying out their roles, and ethical issues include the rights and duties between a company and its employees, customers and suppliers.

Many corporation and professional organisations have a written '*code of ethics*' that defines the professional standards expected of all employees in the company. All employees are expected to adhere to these values whenever they represent the company. The human resource function in a company plays an important role in promoting ethics, and in putting internal HR policies in place relating to the ethical conduct of the employees, as well as addressing discrimination, sexual harassment and ensuring that employees are treated appropriately (including cultural sensitivities in a multicultural business environment).

Companies are expected to behave ethically and not to exploit its workers. There was a case of employee exploitation at the Foxconn plant (an Apple supplier of the *i*Phone) in Shenzhen in China in 2006, where conditions at the plant were so dreadful (long hours, low pay, unreasonable workload and cramped accommodation) that several employees committed suicide. The scandal also raised questions on the extent to which a large corporation such as Apple should protect the safety and well-being of the factory workers of its suppliers. Further, given the profits that Apple makes from the *i*Phone, is it ethical for Apple to allow such workers to be exploited?

Today, the area of *corporate social responsibility* (CSR) has become applicable to the corporate world, and it requires the corporation to be an ethical and responsible citizen in the communities in which it operates (even at a cost to its profits). It is therefore reasonable to expect a responsible corporation to pay its fair share of tax, and to refrain from using tax loopholes to avoid paying billions in taxes on international sales. Today, environment ethics has become topical, and it is concerned with the responsibility of business in protecting the environment in which it operates. It is reasonable to expect a responsible corporation to make the protection of the environment and sustainability part of its business practices.

Unethical business practices refer to those business actions that do not meet the standard of acceptable business operations, and they give the company a bad reputation. It may be that the entire business culture is corrupted or it may be result of the unethical actions of an employee. It is important that such practices be exposed, and this may place an employee in an ethical dilemma (i.e. the loyalty of the employee to the employer versus what is the right thing to do such as exposing the unethical practices).

Some accepted practices in the workplace might cause ethical concerns. For example, in many companies it is normal for the employer to monitor email and Internet use to ensure that employees do not abuse it, and so there may be grounds for privacy concerns. On the one hand, the employer is paying the employee's salary and has a reasonable expectation that the employee does not abuse email and the Internet. On the other hand, the employee has reasonable rights of privacy provided computer resources are not abused.

The nature of privacy is relevant in the business models of several technology companies. For example, Google specialises in Internet-based services and products, and its many products include *Google Search* (the world's largest search engine), *Gmail* for email and *Google Maps* (a web mapping application that offers satellite images and street views). Google's products gather a lot of personal data and create revealing profiles of everyone, which can then be used for commercial purposes.

A Google search leaves traces on both the computer and in records kept by Google, which has raised privacy concerns as such information may be obtained by a forensic examination of the computer, or in records obtained from Google or the Internet Service Providers (ISP). Gmail automatically scans the contents of emails to add context-sensitive advertisements to them and to filter spam, which raises privacy concerns, as it means that all emails sent or received are scanned and read by some computer. Google has argued that the automated scanning of emails is done to enhance the user experience, as it provides customised search results, tailored advertisements, and the prevention of spam and viruses. Google' maps provide location information which may be used for targeted advertisements.

16.3   What Is Computer Ethics?

Computer ethics is a set of principles that guide the behaviour of individuals when using computer resources. Several ethical issues that may arise include intellectual property rights, privacy concerns, as well as the impacts of computer technology on wider society.

The Computer Ethics Institute (CEI) is an American organisation that examines ethical issues that arise in the information technology field. It published the well-known *ten commandments on computer ethics* (Table 16.1) in the early 1990s (Barquin 1992), which attempted to outline principles and standards of behaviour to guide people in the ethical use of computers.

The first commandment says that it is unethical to use a computer to harm another user (e.g. destroy their files or steal their personal data), or to write a program that on execution does so. That is, activities such as spamming, phishing and cyberbullying are unethical. The second commandment is related and may be interpreted that malicious software and viruses that disrupt the functioning of computer systems are unethical. The third commandment says that it is unethical (with some exceptions such as dealing with cybercrime and international terrorism) to read another person's emails, files and personal data as this is an invasion of their privacy.

The fourth commandment argues that the theft or leaking of confidential electronic personal information is unethical (computer technology has made it easier to steal personal information). The fifth commandment states that it is unethical to spread false or incorrect information (e.g. fake news or misinformation spread via email or social media). The sixth commandment states that it is unethical to obtain illegal copies of copyrighted software, as software is considered an artistic or literary work that is subject to copyright. All copies should be obtained legally.

The seventh commandment states that it is unethical to break into a computer system with another user's id and password (without their permission), or to gain unauthorised access to the data on another computer by hacking into the computer system. The eight commandment states that it is unethical to claim ownership of a work that is not yours (e.g. of programs).

The ninth commandment states that it is important for companies and individuals to think about the social impacts of the software that is being created and to create

Table 16.1 Ten commandments on computer ethics

No.	Description
1.	Thou shalt not use a computer to harm other people
2.	Thou shalt not interfere with other people's computer work
3.	Thou shalt not snoop around in other people's computer files
4.	Thou shalt not use a computer to steal
5.	Thou shalt not use a computer to bear false witness
6.	Thou shalt not copy or use proprietary software for which you have not paid
7.	Thou shalt not use other people's computer resources without authorization or proper compensation
8.	Thou shalt not appropriate other people's intellectual output
9.	Thou shalt think about the social consequences of the program you are writing or the system you are designing
10.	Thou shalt always use a computer in ways that ensure consideration and respect for your fellow humans

software only if it is beneficial to society (i.e. it is unethical to create malicious software). The tenth commandment states that communication over computers and the Internet should be courteous, as well as showing respect for others (e.g. no abusive language or spreading false statements).

16.3.1 Ethics and Artificial Intelligence

Weizenbaum created the famous ELIZA program at MIT in 1966, and the program interacted with a user sitting at an electric typewriter, in English, in the manner of a Rogerian psychotherapist.[1]

The name '*Eliza*'[2] was chosen because it could be taught to speak increasingly well, as in Shaw's Pygmalion. It was one of the earliest natural language processing programs, and Weizenbaum gave an example of a conversation between Eliza and a young woman in Weizenbaum (1966). Many users believed that Eliza had real understanding, and they began to unburden themselves in long computer sessions.

The program operated by decomposing the user's input into its constituent parts of speech, and then typing them back at the user in a manner to continue with the dialogue. The program provided very convincing human-like interaction, as may be seen in the sample dialogue below:

User/Patient	Men are all alike.
ELIZA	*In what way?*
User/Patient	They are always bugging us about something or other.
ELIZA	*Can you think of a specific example?*
User/Patient	Well, my boyfriend made me come here.
ELIZA	*Your boyfriend made you come here?*
User/Patient	He says I'm depressed much of the time.
ELIZA	*I am sorry to hear that you are depressed.*
User/Patient	It's true, I am unhappy.
ELIZA	*Do you think that coming here will help you not to be unhappy?*
User/Patient	I need some help, that much seems certain.
ELIZA	*What would it mean to you if you got some help?*
User/Patient	Perhaps, I could learn to get along with my mother.
ELIZA	*Tell me more about your family.*

Weizenbaum was shocked to discover that so many users were convinced that the program had real understanding, and users spent hours sharing their personal problems with the program. It led Weizenbaum to think about the ethics and implications of the artificial intelligence field, and the ability of a relatively trivial

---

[1]Rogerian psychotherapy (person-centred therapy) was developed by Carl Rodgers in the 1940s.

[2]Eliza Doolittle was a working-class character in Shaw's play Pygmalion. She is taught to speak with an upper-class English accent.

program to deceive a naïve user to reveal personal information. He became a leading critic of the AI field and an advocate for professional and social responsibility.

His views on the ethics of AI are discussed in his book 'Computer Power and Human Reason' (Weizenbaum 1976). He displays ambivalence towards computer technology, and he argues that AI is a threat to human dignity, and that AI should not replace humans in positions that require respect and care. He states that machines lack empathy, and that if they replace humans in positions such as police officers or judges, this would lead to alienation and a devaluation of the human condition.

His ELIZA program demonstrated the threat that AI poses to privacy. It is conceivable that an AI program may be developed in the future that is capable of understanding speech and natural languages. Such a program could theoretically eavesdrop on every phone conversation and email, and gather private information on what is said and who is saying it. Such a program could be used by a state to suppress dissent and to eliminate those who pose a threat.

16.3.2   Robots and Ethics

As more and more sophisticated machines and robots are created, it is, of course, essential that intelligent machines behave ethically, and have a moral compass to distinguish right from wrong. It remains an open question as to how to teach a robot right from wrong, and in view of the recent progress that has been made in the AI field, the time is approaching where machines will routinely make ethical decisions.

For example, it is reasonable to expect that driverless cars (self-driving vehicles) will be common on the road in the next 10–20 years. A driverless car is a vehicle that can sense its environment and navigate a route without human intervention. Suppose a self-driving vehicle is travelling on a road and two children roll off a grassy bank on to the road and there is no time for the vehicle to brake. However, if the vehicle swerves to the left, it can avoid the children but hit an oncoming motorbike. *Which decision should the car make and how should it make such a decision?*

This is a variant of the *trolley problem* which is a famous thought experiment in ethics. A train is rushing down a track out of control as its brakes have failed. Disaster lies ahead as five people are tied to the track and will perish in the absence of action. There is sufficient time to flick the points and divert the train down a sidetrack where there is one man tied to the track. Is it ethical to divert the train to do this? Most people would be inclined to take the view that this is the best (least worst) possible outcome.

There is a controversial variant of the problem where the train is rushing towards five people and you are standing on top of a footbridge overlooking the track next to a man with a very bulky rucksack. The only way to save the five people is to push the man to his doom, as his rucksack will block the train and save the five. Is it ethical to deliberately kill or sacrifice another human being to save five others? Most people would say no to this deliberate killing, but it would be valid in the utilitarian school of ethics which seeks to maximise happiness in the world.

Even though the trolley problem is a thought experiment, it is conceivable that a driverless car will face situations where a moral choice must be made (e.g. who to harm or injure such as pedestrians, passengers or driver). Clearly, this raises the importance of the type of ethics that are programmed into the car, and who is to decide what ethics are programmed into a car?

Teaching ethics may involve programming in certain principles, and then the machine learns from scenarios on how to apply the principles to new situations. There is a need for care with machine learning as the machine may learn the wrong lessons, or as its learning evolves it may not be possible to predict its behaviour in the future. Further questions arise as to who is to be held accountable in the event of a machine making incorrect or unethical decisions. For further information on the feasibility of teaching ethics to robots, see the interesting BBC article '*Can we teach robots ethics?*' (BBC Magazine 2017).

16.4   Parnas on Professional Responsibility

Software engineering involves multi-person construction of multi-version programs. It requires the engineer to state precisely the requirements that the software product is to satisfy and to produce designs that will meet these requirements. It involves starting with a precise description of the problem to be solved, producing a design and validating the correctness of the design, and finally, the implementation and testing are performed.

Parnas is a strong advocate of a classical engineering approach, and he argues that computer scientists need the right education to apply scientific and mathematical principles in their work. Software engineers need education on specification, design, turning designs into programs, software inspections and testing. The education should enable the software engineer to produce well-structured programs using module decomposition and information hiding.

Parnas argues that software engineers have individual responsibilities as professionals.[3] They are responsible for designing and implementing high-quality and reliable software that is safe to use. They are also accountable for their own decisions and actions,[4] and have a responsibility to object to decisions that violate professional standards.

---

[3]The concept of accountability for actions dates back thousands of years. The ancient Babylonians employed a code of laws c. 1750 B.C. known as 'The Hammarabi Code'. This included a law that if a house collapsed and killed the owner then the builder of the house would be executed.

[4]However, it is unlikely that an individual programmer would be subject to litigation in the case of a flaw in a program causing damage or loss of life. Most software products are accompanied by a comprehensive disclaimer of responsibility for problems (rather than a guarantee of quality).

Table 16.2 Professional responsibilities of software engineers

No.	Responsibility
1.	Honesty and fairness in dealings with clients
2.	Responsibility for actions
3.	Continuous learning to ensure appropriate knowledge to serve the client effectively

Professional engineers have a duty to their clients to ensure that they are solving the real problem of the client. They need to precisely state the problem before working on its solution. Engineers need to be honest about current capabilities when asked to work on problems that have no appropriate technical solution, rather than accepting a contract for something that cannot be done.[5]

The *licensing of a professional engineer* provides confidence that the engineer has the right education, experience to build safe and reliable products. Otherwise, the profession gets a bad name because of poor work carried out by unqualified people. Professional engineers are required to follow rules of good practice and to object when rules are violated. The licensing of an engineer requires that the engineer completes an accepted engineering course and understands the professional responsibility of an engineer. The professional body is responsible for enforcing standards and certification. The term *'engineer'* is a title that is awarded on merit, but *it also places responsibilities on its holder*.

Engineers have a professional responsibility and are required to behave ethically with their clients. The membership of the professional engineering body requires the member to adhere to the code of ethics of the profession. The code of ethics[6] will detail the ethical behaviour and responsibilities including (Table 16.2).

16.5   ACM Code of Ethics and Professional Conduct

The Association of Computing Machinery (ACM) has defined a code of ethics and professional conduct for its members. The general obligations are detailed in Table 16.3.

---

[5]Parnas applied this professional responsibility faithfully when he argued against the Strategic Defense Initiative (SDI), as he believed that the public (i.e. taxpayers) were being misled and that the goals of the project were not achievable.

[6]These are core values of most mature software companies and many companies today have a code of ethics that employees are required to adhere to.

Table 16.3 ACM code of conduct (general obligations)

No.	Area	Description
1.	Contribute to society and human well-being	Computer professionals must strive to develop computer systems that will be used in socially responsible ways and have minimal negative consequences
2.	Avoid harm to others	Computer professionals must follow best practice to ensure that they develop high-quality systems that are safe for the public. The professional has a responsibility to report any signs of danger in the workplace that could result in serious damage or injury
3.	Be honest and trustworthy	The computer professional will give an honest account of their qualifications and any conflicts of interest. The professional will make accurate statement on the system and the system design, and will exercise care in representing ACM
4.	Be fair and act not to discriminate	Computer professionals are required to ensure that there is no discrimination in the use of computer resources, and that equality, tolerance and respect for others are respected
5.	Respect property rights	The professional must not violate copyright or patent law, and only authorised copies of software should be made
6.	Respect intellectual property	Computer professionals are required to protect the integrity of intellectual property, and must not take credit for another person's ideas or work
7.	Respect the privacy of others	The professional must ensure that any personal information gathered for a specific purpose is not used for another purpose without the consent of the individuals. User data observed during normal system operation must be treated with the strictest confidentiality
8.	Respect confidentiality	The professional will respect all confidentiality obligations to employers, clients and users

16.6 British Computer Society Code of Conduct

The BCS has a code of conduct that defines the standards expected of BCS members, and it applies to all grades of members during their professional work. Any known breaches of the BCS codes by a member are investigated by the BCS, and appropriate disciplinary procedures followed. The main parts of the BCS code of conduct are listed in Table 16.4.

Table 16.4 BCS code of conduct

Area	Description
Public interest	Due regards to rights of third parties Conduct professional activities without discrimination Promote equal access to IT
Professional competence and integrity	Only do work within professional competence Do not claim competence that you do not possess Ongoing development of knowledge/skills Avoid injuring others Reject bribery and unethical behaviour
Duty to relevant authority	Carry out professional responsibilities with due care and diligence Exercise professional judgment Accept professional responsibility for work
Duty to the profession	Uphold reputation of profession and BCS Seek to improve professional standards Act with integrity Support other members in their professional development

16.7 Review Questions

1. What is ethics?
2. Describe the main schools of ethics.
3. What is business ethics?
4. Give examples of unethical behaviour.
5. Discuss the relevance of the Eliza program to computer ethics.
6. Describe Parnas's contributions to the debate concerning the professional responsibility of software engineers.
7. Describe the ACM code of ethics and professional conduct.
8. Describe the BCS code of conduct.

16.8 Summary

Ethics is a branch of philosophy that deals with moral questions such as what is right or wrong, and the right behaviour for an individual in a given situation. There are various schools of ethics such as the relativist position, cultural relativism, deontological ethics and utilitarianism.

Business ethics (also called corporate ethics) is concerned with ethical principles and moral problems that arise in a business environment. They refer to the core principles and values of the organisation, and apply throughout the organisation. The ethical issues include the rights and duties between a company and its employees, customers and suppliers.

Professional ethics are a code of conduct that governs how members of a profession deal with each other and with third parties. It defines the fundamental principles of the organisation, which is an indication of the professionalism of the organisation.

Several organisations such as the ACM and BCS have developed a code of conduct for their members, and violations of the code by members are subject to investigations and disciplinary procedures.

Chapter 17
Innovation in the Computing Field

Key Topics

Distributed system
Service-oriented architecture
Software as a Service
Cloud computing
Aspect-oriented software engineering
Embedded systems
Innovation in software engineering

17.1 Introduction

The process of translating a business idea or invention into a product or service that adds value and that people will pay for is termed *innovation*. However, for a business idea to be termed innovative, it must be commercially viable at an economic cost that people will be willing to pay, and it must satisfy a specific customer need (as otherwise there will be no demand for it).

There are two broad categories of innovations namely *evolutionary* and *revolutionary* innovations. An evolutionary innovation is generally brought about by incremental advances in technology, whereas a revolutionary innovation is often totally new and completely different from the existing products in the market place (e.g. the development of the Apple Macintosh or iPhone were paradigm shifts from the existing state of the art). There is generally greater risk with a revolutionary innovation as it is creating an entirely new product, whereas evolutionary innovations generally involve less risk.

The success of hi-tech companies relies on the creativity and innovation of its staff, and therefore it is important to foster innovation in the workplace. An innovative work environment generally has a low power distance between management and staff, with an emphasis on open communication and inter-department collaboration. Brainstorming sessions to come up with innovative ideas or solutions to problems are encouraged, as well as the use of a suggestion box where employees can submit ideas or improvement suggestions as well as making them to their supervisor.

© Springer International Publishing AG, part of Springer Nature 2018
G. O'Regan, *World of Computing*,
https://doi.org/10.1007/978-3-319-75844-2_17

The objective of this chapter is to give a flavour of several topics that have become relevant to the computing field in recent times. The software field is highly innovative and is continually evolving, and this has led to the development of many new technologies and systems. This includes distributed systems, service-oriented architecture (SOA), Software as a Service (SaaS), cloud computing, embedded systems, and many more. Software engineering needs to continually respond to the emerging technology trends with innovative solutions and methodologies to support the latest developments.

A distributed system is a collection of computers that appears to be a single system, and many large computer systems used today are distributed systems. A distributed system allows hardware and software resources to be shared, and it supports concurrency with multiple processors running on different computers on the network.

SOA is a way of developing a distributed system consisting of stand-alone web services that may be executing on distributed computers in different geographic regions. SaaS allows software to be hosted remotely on a server (or servers), and the user can access the software over the Internet through a web browser. Cloud computing is a type of internet-based computing that provides computing resources and various other services on demand.

An embedded system is a computer system within a larger electrical or mechanical system, and it is *embedded* as part of a complete system that includes hardware and mechanical parts. An embedded system is usually designed to do a specific task rather than as a general-purpose device, and it may be subject to real-time performance constraints.

17.2   Distributed Systems

A *distributed system* (Fig. 17.1) is a collection of computers, interconnected via a network, which is capable of collaborating on a task. It appears to be a single integrated computing system to the user, and most large computer systems today are distributed systems. The components (or nodes) of a distributed system are located on networked computers, and interact to achieve a common goal.

The communication and coordination of action is via message passing. A distributed system is not centrally controlled, and as a result the individual computers may behave differently at different times, and each computer has a limited and incomplete view of the system.

A distributed system allows hardware and software resources (e.g. printers and files) to be shared, and information may be shared between people and processes located in distant geographical regions. It supports concurrency with multiple processors running on different computers on the network. The processors in a distributed system run concurrently in parallel, and each computer is running on its own local operating system.

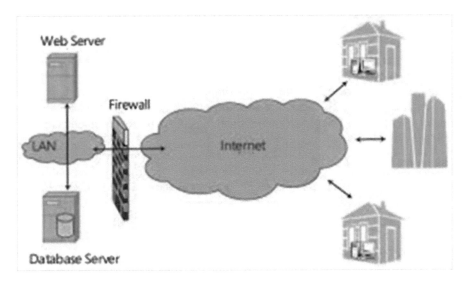

Fig. 17.1 A distributed system

A distributed system is designed to tolerate failures on individual computers, and the system is designed to be reliable and to continue service when a node fails. That is, a distributed system needs to be designed to be fault tolerant, and it must remain available if there are hardware, software or network failures. This requires building in redundancy and recovery features (e.g. duplicating information on several computers). The fault tolerant design allows continuity of service (possibly a degraded service) when failures occur.

The design of distributed systems is more complex than a centralised system, as there may be complex interactions between its components and the system infrastructure. The performance of the distributed system is dependent on the network bandwidth and load, as well as on the speed of the computers that are on the network. This differs from a centralised system, which is dependent on the speed of a single processor. The performance and response time of a distributed system may vary (and be unpredictable) depending on the network load and network bandwidth, and so the response time may vary from user to user.

The nodes in a distributed system are often independent systems with no central control, and the network connecting the nodes is a complex system, which is not controlled by the systems using the network. There are many applications of distributed system in the telecommunication domain, such as fixed line, mobile and wireless networks, company intranets, the Internet and the World Wide Web. Next, we describe SOA and how it is used in distributed systems.

17.3   Service-Oriented Architecture

The objective of this section is to give a brief introduction to *service-oriented architecture* (SOA), which is a way of developing a distributed system using stand-alone web services executing on distributed computers in different geographic regions. It is an approach to creating an architecture based upon the use of services, where a service may carry out some small function such as producing data or validating a customer.

A web service is a computational or information resource that may be used by another program, and it allows a *service provider* to provide a service to an application (*service requestor*) that wishes to use the service. The web service may be accessed remotely, and is acted upon independently. The service provider is responsible for designing and implementing the service, and specifying the interface to the service.

The service is platform and implementation language independent, and it is designed and implemented by the service provider with the interface to the service specified. Information about the service is published in an accessible *service registry*, and service clients (requestors) can locate the service provider and link their application with the specific service and communicate with it. The idea of a SOA is illustrated in Fig. 17.2.

There are several standards that support communication between services, as well as standards for service interface definition. These are discussed in Sommerville (2010).

17.4   Software as a Service

The idea of *Software as a Service* (SaaS) is that the software may be hosted remotely on a server (or servers), and access provided to it over the Internet through a web browser. The functionality is provided at the remote server with client access provided through the web browser.

Fig. 17.2 Service-oriented architecture

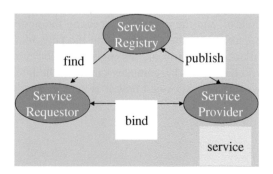

The cost model for traditional software is made up of an upfront cost for a perpetual license and optional ongoing support fees. SaaS is a software licensing and delivery model where the software is licensed to the user on a subscription basis. The software provider owns and provides the service, whereas the software organisation that is using the service will pay a subscription for its use. Occasionally, the software is free to use with funding for the service provided through advertisements, or there may be a free basic service provided with charges applied for the more advanced version.

A key benefit of SaaS is that the cost of hosting and management of the service is transferred to the service provider, with the provider responsible for resolving defects and installing upgrades of the software. Consequently, the initial set-up costs for users are significantly less than for traditional software.

The disadvantages to the user are that data must be transferred at the speed of the network, and the transfer of a large amount of data may take a lot of time. The subscription charges may be monthly or annual, with extra charges possibly due depending on the amount of data transferred.

17.5 Cloud Computing

Cloud computing is a type of Internet-based computing that provides computing processing resources on demand. It provides access to a shared pool of configurable computing resources such as networks, servers and applications on demand, and such resources may be provided and released with minimal effort. It provides users and organisations with capabilities to store and process their data at third-party data centres that may be in distant geographical locations.

A key advantage of cloud computing is that it allows companies to avoid large upfront infrastructure costs such as purchasing hardware and servers, and it allows organisations to focus on their core business. Further, it allows companies to get their applications operational in a shorter space of time, as well as providing an efficient way for companies to adjust resources to deal with fluctuating demand. Companies can scale up as computing needs increase and scale down as demand decreases. Cloud providers generally use a '*pay as you go*' model (Fig. 17.3).

Among the well-known cloud computing platforms are Amazon's Elastic Compute Cloud, Microsoft's Azure and Oracle's cloud. The main enabling technology for cloud computing is virtualisation, which separates a physical computing device into one or more virtual devices. Each of the virtual devices may be easily used and managed to perform computing tasks, and this leads to the creation of a scalable system of multiple independent computing devices that allows the idle physical resources to be allocated and used more effectively.

Cloud computing providers offer their services according to different models. These include *infrastructure as a service* (IaaS) where computing infrastructure

Fig. 17.3 Cloud computing.
Creative commons

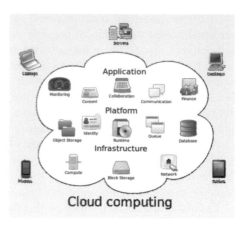

such as virtual machines and other resources are provided as a service to sub-scribers. *Platform as a service* (Paas) provides capability to the consumer to deploy infrastructure related or application related that are supported by the provider onto the cloud. PaaS vendors offer a development platform to application developers. *Software as a Service* (SaaS) provides capability to the consumer to use the pro-vider's applications running on a cloud infrastructure through a web browser or a program interface. Cloud providers manage the infrastructure and platforms that run the applications.

17.6   Embedded Systems

An embedded system is a computer system within a larger electrical or mechanical system that is usually subject to real-time constraints. The computer system is *embedded* as part of a complete system that includes hardware and mechanical parts. Embedded systems vary from personal devices such as MP3 players and mobile phones, to household devices such as dishwashers and cookers, to the automotive sector and to traffic lights. An embedded system is usually designed to do a specific task rather than as a general-purpose device, and it may be subject to real-time performance constraints (Fig. 17.4).

Some embedded systems are termed *reactive systems* as they react to events that occur in their environment, and so their design is often based on a stimulus-response model. An event (or condition) that occurs in the system envi-ronment that causes the system to respond in some way is termed a stimulus, and a response is a signal sent by the software to its environment. For example, in the automotive sector there are sensors in a car that detect when the temperature in the engine goes too high, and the response may be an audio alarm and visual warning to the driver.

Fig. 17.4 Example of an embedded system

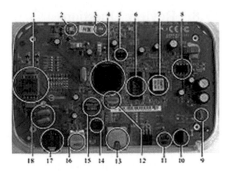

One of the earliest embedded system was the guidance computer developed for the Minuteman II missile (O' Regan 2016) in the mid-1960s. Embedded systems are ubiquitous today, and they control many devices that are in common use such as microwave ovens, washing machines, coffee makers, clocks, DVD players, mobile phones and televisions.

Embedded systems became more popular following the introduction of the microprocessor in the early 1970s, as cheap microprocessors could fulfil the same role as many components. Most microprocessors produced today are used as components of embedded systems.

17.7 Software Engineering and Innovation

The software field is highly innovative and many new technologies and systems have been developed. We have discussed a sample of these innovations in this chapter, and the software engineering field needs to continually respond to emerging technology trends with innovative solutions and methodologies.

There have been many innovations in software engineering since its birth in the late 1960s. These include the waterfall and spiral lifecycle models, the Rational Unified Process; the Agile methodology; software inspections and reviews; software testing and test-driven development; information hiding, object-oriented design and development; formal methods and UML; software process improvement, the CMM, CMMI and ISO SPICE.

There is also the need to focus on best practice in software engineering, as well as emerging technologies from various research programs. Piloting or technology transfer of innovative technology is an important part of continuous improvement.

17.8 Review Questions

1. What is a distributed system?
2. What is service-oriented architecture?
3. What is Software as a Service?
4. What is cloud computing?
5. What is embedded software engineering?
6. Describe the various models that are used in cloud computing.
7. Describe various innovations in software engineering.

17.9 Summary

The process of translating a business idea or invention into a product or service that adds value and that people will pay for is termed innovation, and the success of a business relies on the creativity and innovation of its staff. It is therefore important to foster innovation in the workplace, and an innovative work environment generally has an emphasis on open communication and inter-department collaboration. Brainstorming sessions to come up with innovative ideas or solutions to problems are encouraged.

This chapter gave a brief introduction to distributed systems, SOA, SaaS, cloud computing, embedded systems and aspect-oriented software engineering.

A distributed system is a collection of interconnected computers that appears to be a single system. SOA is a way of developing a distributed system consisting of stand-alone web service executing on distributed computers in different geographic regions. SaaS allows software to be hosted remotely on a server (or servers), and access is provided to it over the Internet through a web browser. Cloud computing is a type of internet-based computing that provides computing resources and various other services on demand.

An embedded system is a computer system within a larger electrical or mechanical system, and it is usually designed to do a specific task rather than as a general-purpose device, and it may be subject to real-time performance constraints.

Chapter 18
Epilogue

We embarked on a long journey is this book to provide readers with a concise introduction to the world of computing. The first chapter introduced analog and digital computers, and the von Neumann architecture which is the fundamental architecture underlying a digital computer.

We then discussed the foundations of computing including the binary number system and the Step Reckoner calculating machine. Babbage designed the Difference Engine as a machine to evaluate polynomials, and his Analytic Engine provided the vision of a modern computer. Boole's symbolic logic provided the foundation for digital computing.

We presented a brief history of computing including the first digital computers, the first commercial computers, the SAGE air defence system, the invention of the transistor at Bell Labs, the invention of the integrated circuit at Texas Instruments, the development of the IBM System/360 and later mainframes and minicomputers. We discussed the invention of the microprocessor, and how it led to home computers. We discussed the introduction of the IBM personal computer.

We introduced the fundamental mathematics used in computing, including sets, relations and functions. Sets are collections of well-defined objects; relations indicate relationships between members of two sets A and B; and functions are a special type of relation where there is exactly (or at most) one relationship for each element $a \in A$ with an element in B.

We then presented a short introduction to algorithms, where an algorithm is a well-defined procedure for solving a problem. It consists of a sequence of steps that takes a set of values as input, and produces a value (or set of values) as output. It is an exact specification of how to solve the problem, and it may be implemented by a computer program.

We presented an introduction to logic for computing, including propositional and predicate logic. Propositional logic is the study of propositions, where a proposition is a statement that is either true or false. It may be used to encode simple arguments that are expressed in natural language, and to determine their

© Springer International Publishing AG, part of Springer Nature 2018
G. O'Regan, *World of Computing*,
https://doi.org/10.1007/978-3-319-75844-2_18

validity. Predicate logic allows complex facts about the world to be represented, and new facts may be determined via deductive reasoning.

We then discussed human–computer interaction (HCI), which is a branch of computer science that is concerned with the design, evaluation and implementation of interactive computing systems for human use. It is focused on the interfaces between people and computers.

We presented a short introduction to programming languages starting with machine languages; to assembly languages; to early high-level procedural languages such as Fortran and COBOL; to later high-level procedural languages such as Pascal and C and to object-oriented languages such as C++ and Java. Functional programming languages and logic programming languages were discussed, and there was a short discussion on syntax and semantics.

We then presented a short introduction of the software engineering field, including the waterfall and spiral lifecycles, the rational unified process and the popular Agile methodology. We discussed the key activities in the waterfall model such as requirements, design, implementation, unit, system and acceptance testing.

We gave a short introduction to operating systems including the IBM OS/360, the MVS and VM operating systems, the UNIX operating system which is a multi-user and multi-tasking operating system and the DEC VAX/VMS operating system. Microsoft developed MS/DOS operating system for IBM compatible personal computers, and its Windows operating system was a response to the GUI-driven Apple Macintosh. We discussed Android and iOS, which are popular operating systems for mobile devices.

We then presented a short introduction to databases including the hierarchical and network models. We discussed the relational model as developed by Codd at IBM in more detail, and most databases used today are relational.

We gave a short introduction to telecommunications, including the AXE system (the first fully automated digital switching system), and the development of mobile phone technology. We discussed the introduction of the first mobile phone, the DynaTAC, by Motorola.

We described the Internet revolution starting from ARPANET, which was a packet switched network and TCP/IP, which is a set of network standards for interconnecting networks and computers. These developments led to the birth of the Internet and the World Wide Web. We discussed applications of the World Wide Web, and the dot-com bubble and burst of the late 1990s/early 2000. We discussed the Internet of Things and the Internet of Money.

We discussed the invention of the smartphone and the rise of social media, and the impact of Facebook and Twitter in social networking. Facebook has become a way for young people to discuss their hopes and aspirations as well as a tool for social protest and revolution. Twitter has become a popular tool in political communication, and it is also an effective way for businesses to advertise its brand to its target audience. We discussed how social media has facilitated the growth of fake news, and the challenges that poses to western society.

We discussed legal aspects of computing including the overlap of the law and computing. This covered the legal aspects of digital information and software, as well as the legal aspects of the Internet. We discussed intellectual property law including patents, copyright, trademarks and trade secrets, and the problem of hacking where a hacker is a person who uses his computer skills to gain unauthorised access to a computer system.

A software license is a legal agreement between the copyright owner and the licensee, which governs the use or distribution of software to the user. The two most common categories of software licenses that may be granted under copyright law are those for proprietary software and those for free open source software. We discussed the legal aspects of bespoke software development, and a legal contract is prepared between the supplier and the customer.

We discussed ethics and professional responsibility in computing. Ethics is a branch of philosophy that deals with moral questions such as what is right or wrong, professional ethics are a code of conduct that governs how members of a profession should deal with each other and with third parties.

We discussed innovation in the computer field including distributed systems, service-oriented architecture (SOA), Software as a Service (SaaS), cloud computing and embedded systems. Finally, we summarised the journey that we have travelled in the book.

Glossary

ABC Atanasoff-Berry Computer

ACM Association Computing Machinery

AI Artificial Intelligence

ALGOL Algorithmic Language

AMPS Advanced Mobile Phone System

ANS Advanced Network Services

ANSI American National Standards Institute

API Application Programmer Interface

ARPA Advanced Research Projects Agency

ASCC Automatic Sequence Controlled Calculator

ASCII American Standard Code for Information Interchange

AXE Automatic Exchange Electric switching system

B2B Business to Business

B2C Business to Consumer

BASIC Beginners All-purpose Symbolic Instruction Code

BBN Bolt, Beranek and Newman

BCS British Computer Society

BIOS Basic Input Output System

BNF Backus Naur Form

CD Compact Disc

© Springer International Publishing AG, part of Springer Nature 2018
G. O'Regan, *World of Computing*,
https://doi.org/10.1007/978-3-319-75844-2

CDC Control Data Corporation

CDMA Code Division Multiple Access

CEI Computer Ethics Institute

CEO Chief Executive Officer

CERN Conseil Européen pour la Recherche Nucléaire

CERT Certified Emergency Response Team

CMM® Capability Maturity Model

CMMI® Capability Maturity Model Integration

CMS Conversational Management System

COBOL Common Business Oriented Language

CODASYL Conference on Data Systems Languages

COPQ Cost of Poor Quality

COTS Customised Off The Shelf

CP Control Program

CP/M Control Program for Microcomputers

CPU Central Processing Unit

CRT Cathode Ray Tube

CSR Corporate Social Responsibility

CTSS Compatible Time-Sharing System

DARPA Defense Advanced Research Project Agency

DB Database

DBA Database Administrator

DBMS Database Management System

DDL Data Definition Language

DEC Digital Equipment Corporation

DL/1 Data Language 1

DML Data Manipulation Language

DNS Domain Naming System

DOS Disk Operating System

DoS Denial of Service

DRI Digital Research Incorporated

DSDM Dynamic Systems Development Method

DVD Digital Versatile Disc

EDVAC Electronic Discrete Variable Automatic Computer

ENIAC Electronic Numerical Integrator and Computer

ESA European Space Agency

ETACS Extended TACs

ETSI European Telecommunications Standards Institute

EULA End-user license agreement

FAA Federal Aviation Authority

FDMA Frequency Division Multiple Access

FTP File Transfer Protocol

FORTRAN Formula Translation

FOSS Free Open Source Software

FSM Finite State Machine

GB Giga Byte

GECOS General Electric Comprehensive Operating System

GFW Great Firewall of China

GL Generation Language

GPL GNU Public License

GPRS General Packet Radio Service

GSM Global System Mobile

GUAM Generalised Update Access Method

GUI Graphical User Interface

HCI Human Computer Interaction

HP Hewlett Packard

HR Human Resources

HTML Hypertext Markup Language

HTTP Hyper Text Transport Protocol

IBM International Business Machines

IC Integrated Circuit

IDMS Integrated Database Management System

IDS Integrated Data Store

IE Internet Explorer

IEC International Electrotechnical Commission

IEEE Institute of Electrical and Electronic Engineers

IMP Interface Message Processor

IMS Information Management System

INWG International Network-Working Group

IP Internet Protocol

IPCS Interactive Problem Control System

IPO Initial Public Offering

ISEB Information Systems Examination Board

ISO International Standards Organization

ISP Internet Service Provider

IT Information Technology

JAD Joint Application Development

JCL Job Control Language

JVM Java Virtual Machine

KB Kilo Byte

KLOC Thousand Lines of Code

LD Limited Domain

LED Light Emitting Diode

LEO Lyons Electronic Office

LEO Low Earth Orbit

LISP List Processor

LSI Large Scale Integration

MADC Manchester Automatic Digital Computer

MB Mega Byte

ME Millennium

MFT Multiple Programming with a Fixed number of Tasks

MIPS Million Instructions Per Second

MIT Massachusetts Institute of Technology

MITS Micro Instrumentation and Telemetry System

MOS Metal Oxide Semiconductor

MSI Medium Scale Integration

MS/DOS Microsoft Disk Operating System

MTX Mobile Telephone Exchange

MVS Multiple Virtual Storage

MVT Multiple Programming with a Variable number of Tasks

NAP Network Access Point

NASA National Aeronautics and Space Administration

NATO North Atlantic Treaty Organisation

NCP Network Control Protocol

NLS On Line System

NMT Nordic Mobile Telephony system

NORAD North American Aerospace Defence

NPL National Physical Laboratory

NSF National Science Foundation

OS Operating System

OSS Open Source Software

PARC Palo Alto Research Centre

PC Personal Computer

PCB Process Control Block

PCP Principal Control Program

PC/DOS Personal Computer Disk Operating System

PET Personal Electronic Transactor

PDA Personal Digital Assistant

PDP Programmed Data Processor

PL/M Programming Language for Microcomputers

PTT Postal Telephone and Telegraph

RAD Rapid Application Development

RAM Random Access Memory

RDBMS Relational Database Management System

RIM Research in Motion

ROM Read Only Memory

RSA Rivest, Shamir and Adleman

RSCS Remote Spooling Communications Subsystem

RUP Rational Unified Process

SAGE Semi-Automatic Ground Environment

SCAMPI Standard CMMI Appraisal Method for Process Improvement

SECD Stack, Environment, Control, Dump

SEI Software Engineering Institute

SID Sound Interface Device

SILK Speech, Images, Language, Knowledge

SIM Subscriber Identity Module

SLA Service Level Agreement

SM Service Mark

SMS Short Message Service

SMTP Simple Mail Transfer Program

SNS Social Networking Site

SOA Service Oriented Architecture

SOW Statement of Work

SPREAD System Programming, Research, Engineering and Design

SQL Structured Query Language

SRI Stanford Research Institute

SSEC Selective Sequence Electronic Computer

SSI Small Scale Integration

SSL Secure Socket Layer

TACS Total Access Communication

TCP Transport Control Protocol

TM Trade Mark

TSO Time Sharing Option

UAT User Acceptance Testing

UCD User-centred design

UCLA University of California (Los Angeles)

UDP User Datagram Protocol

ULSI Ultra Large-Scale Integration

UML Unified Modelling Language

UNIVAC Universal Automatic Computer

URL Universal Resource Locator

VAX Virtual Address eXtension

VDM Vienna Development Method

VLSI Very Large-Scale Integration

VM Virtual Memory

VMS Virtual Memory System

VUI Voice User Interface

WCDMA Wideband CDMA

WIMP Windows, Icons, Menus, Pointers

References

Ackrill JL (1994) Aristotle the Philosopher. Clarendon Press Oxford, Oxford

Anderson T, Dahlin M (2014) Operating systems: principles and practice. Recursive Books

Bagnall B (2012) Commodore: a company on the edge, 2nd edn. Variant Press

Barquin RC (1992) In Pursuit of a 'Ten Commandments' for Computer Ethics. Computer Ethics Institute

BBC Magazine (2017) Can we teach Robots Ethics. BBC Magazine, 17 Oct 2017

Berners-Lee T (2000) Weaving the Web. Collins Book, New York

Bloomberg Business Week Magazine (2004) The Man Who Could Have Been Bill Gates, Oct 2004

Boehm B (1988) A Spiral Model for software development and enhancement. Computer, May 1988

Boole G (1848) The calculus of logic. Cambridge and Dublin Mathematical Journal, vol III, pp 183–198

Boole G (1958) An investigation into the laws of thought. Dover Publications (First published in 1854)

Brooks F (1975) The Mythical Man Month. Addison Wesley, Boston

Brooks F (1986) No Silver Bullet. Essence and accidents of software engineering. Information processing. Elsevier. Amsterdam

Bush V (1945) As we may think. The Atlantic Monthly, 176, No. 1, July 1945

Buxton IN, Naur P, Randell B (1975) Software engineering. Petrocelli. Report on two NATO Conferences held in Garmisch, Germany (October1968) and Rome, Italy (October 1969)

Chaum D (1982) Blind signatures for untraceable payments. Advances in cryptology. Proceedings of crypto, vol 82(3), pp 199–203

Chesbrough H, Rosenbloom R (2002) The role of the business model in capturing value from innovation: evidence from Xerox Corporation's technology spin-off companies. Ind Corp Change 11(3):529–555

Chrissis MB, Conrad M, Shrum S (2011) CMMI. Guidelines for process integration and product improvement, 3rd edn. SEI series in software engineering. Addison Wesley, Boston

Codd EF (1970) A relational model of data for large shared data banks. Commun ACM 13(6):377–387

Date CJ (1981) An introduction to database systems, 3rd edn. The systems programming series

Deitel HM (1990) Operating systems, 2nd edn. Addison Wesley, Boston

Dijkstra EW (1968) Go to statement considered harmful. Commun ACM, Mar 1968

Dijkstra EW (1972) Structured programming. Academic Press, London

Dijkstra EW (1976) A discipline of programming. Prentice Hall, Englewood Cliffs

Fagan M (1976) Design and code inspections to reduce errors in software development. IBM Syst J 15(3)

Gertner J (2013) The idea factory: Bell Labs and the great age of American innovation. Penguin Books, New York

Gilb T, Graham D (1994) Software inspections. Addison Wesley, Boston

© Springer International Publishing AG, part of Springer Nature 2018
G. O'Regan, *World of Computing*,
https://doi.org/10.1007/978-3-319-75844-2

Greenfield A (2017) Rise of the machines. Who is the internet of things good for? Guardian Article. 6 June 2017. https://www.theguardian.com/technology/2017/jun/06/internet-of-things-smart-home-smart-city.

Jacobson I et al (2005) The unified modelling language, User Guide, 2nd edn. Addison Wesley Professional, Boston

Jacobson I, Booch G, Rumbaugh J (1999) The unified software development process. Addison Wesley, Boston

Kahn B, Cerf V (1974) Protocol for packet network interconnections. IEEE Trans Commun Technol

Kelly J (1997) The essence of logic. Prentice Hall, Englewood Cliffs

Kernighan B (1981) Why Pascal is not my favourite language. AT&T Bell Laboratories

Kernighan B, Ritchie D, The C Programming Language, 1st edn

Leibniz WG (1703) Explication de l'Arithmétique Binaire. Memoires de l'Academie Royale des Sciences

Levy S (1984) Hackers: heroes of the computer revolution. O'Reilly Media

Licklider JCR (1960) Man-computer symbiosis. IRE Trans Hum Factors Electron HFE 1:4–11

MacHale D (1985) Boole. Cork University Press, Cork

Malmsten E, Portanger E (2002) Boo Hoo. $135 Million, 18 Months. . . A Dot.Com Story from Concept to Catastrophe. Arrow

Menabrea LF (1842a) Sketch of the analytic engine invented by Charles Babbage (trans: Lady Ada Augusta, Countess of Lovelace). Bibliothèque Universelle de Genève, Oct 1842, No. 82

Menabrea LF (1842b) Sketch of the analytical engine invented by Charles Babbage (trans: Lovelace LA). Bibliothèque Universelle de Genève

Meurling J, Jeans R (2001) The Ericsson Chronicle: 125 Years in Telecommunications. Informationsforlaget, Stockholm

Moore G (1965) Cramming more components onto integrated circuits. Electronics Magazine

Motorola Museum of Electrics and Motorola (1999) Motorola (CB)—a journey through time and technology. Purdue University Press, West Lafayette

Nakamoto S (2008) Bitcoin. A peer-to-peer electronic cash system

Naur P (ed) (1960) Report on the algorithmic language: ALGOL 60. Commun ACM 3(5):299–314

O' Regan G (2010) Introduction to software process improvement. Springer, New York

O' Regan G (2012) Mathematics in computing. Springer, New York

O' Regan G (2013) Giants of computing. Springer, New York

O' Regan G (2014) Introduction to software quality. Springer, New York

O' Regan G (2015) Pillars of computing. Springer, New York

O' Regan G (2016) Introduction to the history of computing. Springer, New York

O' Regan G (2017a) Guide to discrete mathematics. Springer, New York

O' Regan G (2017b) Concise guide to software engineering. Springer, New York

O' Regan G (2017c) Concise guide to formal methods. Springer, New York

Office of Government Commerce (2004) Managing Successful Projects with PRINCE2

Parnas D (1972) On the criteria to be used in decomposing systems into modules. Commun ACM 15(12)

Parnas D (2001) Software fundamentals. Collected papers (Weiss D, Hoffman D, eds). Addison Wesley, Boston

Piff M (1991) Discrete mathematics. An introduction for software engineers. Cambridge University Press, Cambridge

Plotkin G (1981) A structural approach to operational semantics. Technical report DAIM FN-19. Computer Science Department, Aarhus University, Denmark

Porter ME (1998) Competitive advantage. Creating and sustaining superior performance. Free Press, New York

Pugh EW (2009) Building IBM: shaping an industry and its technology. MIT Press, Cambridge

Robbins A (2005) Unix in a Nutshell, 4th edn. O'Reilly Media

Royce W (1970) The software lifecycle model (Waterfall Model). In: Proc. WESTCON, August 1970

Schaefer MW (2014) The Tao of Twitter. Changing your life and business 140 characters at a time, 2nd ed. McGraw-Hill, New York

Shannon C (1937) A symbolic analysis of relay and switching circuits. Masters thesis, Massachusetts Institute of Technology

Shockley W (1950) Electrons and holes in semiconductors with applications to transistor electronics. Van Nostrand, New York

Sir Thomas Heath (1956) Euclid. The thirteen books of the elements, vol 1 (trans: Sir Thomas Heath). Dover Publications, New York (First published in 1925)

Shneiderman B, Plaisant C (2005) Designing the user interface. Pearson Education, London

Sommerville I (2010) Software engineering, vol 9. Pearson, London

Spivey JM (1992) The Z Notation. A reference manual. Prentice Hall International Series in Computer Science

Standard CMMI Appraisal Method for Process Improvement. CMU/SEI-2006-HB-002. V1.2, Aug 2006

Turner D (1985) Miranda. In: Proceedings IFIP Conference, Nancy France, Springer LNCS (201), Sept 1985

von Neumann J (1945) First draft of a report on the EDVAC. University of Pennsylvania

Weizenbaum J (1966) ELIZA. A computer program for the study of natural language communication between man and machine. Commun ACM 9(1):36–45

Weizenbaum J (1976) Computer power and human reason: from judgment to calculation. W.H. Freeman and Co, San Francisco

Index

© Springer International Publishing AG, part of Springer Nature 2018
G. O'Regan, *World of Computing*,
https://doi.org/10.1007/978-3-319-75844-2